The Anonymous Us Project

Volume 1

Compiled and edited by Alana S. Newman

BROADWAY PUBLICATIONS

Published in the United States of America by
Broadway Publications,
1841 Broadway, Suite 211,
New York, NY 10023.

ISBN 978-1-105-93678-4

Copyediting and book design by Betsy Wells Stokes,
www.BetsyWellsStokes.com.

This project was made possible by the support of
the Institute for American Values (IAV), www.AmericanValues.org.
The opinions expressed here, however, are those of the author and contributors alone.
IAV retains no control over the project's content or direction.

All personal names in contributors' accounts have been changed.

Acknowledgments

Whether you believe it takes a village to raise a family or a family to raise a village, it most certainly takes coordination and teamwork to create a successful and functional web initiative and then turn it into a book.

Thank you to Elizabeth Marquardt and others at the Center for Marriage and Families, for listening to me, investing in me, and guiding me through the process of turning the Anonymous Us Project into a reality. Without you, this project would be a dead idea in the pages of a forgotten diary.

Thank you to Alma Phipps for your incredible logo design and Josephine Tramontano and Lisa Webb, who built AnonymousUs.org from the ground up. You made the site wonderfully user-friendly, and I am so proud.

Thank you to my husband for your never-ending support of Anonymous Us. You are an amazing husband and father. And thank you for having such high standards for yourself and your family.

Finally, I want to thank my functioning family, especially my mom, for enduring the pain of where my journey has taken us all and trying your best to understand why this is important.

Mission of the Anonymous Us Project

The Anonymous Us Project acknowledges that many members of the 3PR (third-party reproduction) community have serious opinions and experiences regarding these alternative family structures that they'd like to add to the public discussion but feel hesitant to speak publicly on due to privacy concerns. This project gives them an opportunity to be heard without revealing their identities or risking hurt to their loved ones.

The Anonymous Us Project believes that while anonymity in reproduction hides the truth, anonymity in storytelling helps reveal it.

We hope this project will fill out the conversation on 3PR. We hope it will inspire more truth and transparency. We hope it will guide researchers, scholars, and policy makers in asking the right questions and considering the full effects of these practices. We hope it will help shape healthier families and happier people.

There is no greater agony than
bearing an untold story inside you.

—*Maya Angelou*

Contents

Introduction

I began the Anonymous Us Project after I discovered that many of my donor-conceived peers were unwilling to give their testimonials publicly because of real or perceived fears—mainly of hurting the feelings of their family members. This barrier was preventing their participation in public discourse on the topic and leading to a serious vacuum in insights and intellectual contributions to the debate. I thought, *How can we as a nation make wise decisions about family structure, third-party reproduction, and gamete donation without the participation of and insights from those who have been most directly affected by these practices?*

Although I myself am donor-conceived and built AnonymousUs.org primarily as a tool for other donor-conceived people, it is a platform for all interested voices, and I encourage everyone involved with reproductive technologies, as well as adoption, to share insights and revelations there.

As the founder of the Anonymous Us Project, I find it important that contributors feel secure that their perspectives are respectfully processed and graciously received. I do not want to speak for anyone else, but I do see my role as a microphone. I'm interested in the big ideas experienced in the process and in understanding people's motivations. I want to understand the phenomenon of reproductive technologies from every angle, beginning with the lived experiences and personal views of the people involved.

In part two, I hope you'll allow me to serve you a few concepts that seem to be recurring concerns for those affected by 3PR (third-party reproduction). These are not comprehensive. Nor are they arranged in any particular order. I have no information regarding how common these opinions and circumstances are, but I propose them as potential research topics.

In no way am I intending to alienate those who are grateful for the reproductive technologies that are available to them and that have improved their lives. In addition, my research is not a statement on women's rights, abortion, same-sex unions, or any other activity in which willing adults engage—except in that it involves the children-become-adults who experience it along with them.

I first began writing about my experiences as a donor-conceived woman through guest blogging at FamilyScholars.org. The editor of the blog, Elizabeth Marquardt, was releasing a report titled "My Daddy's Name Is Donor," and she was looking for writers who could comment on the topic. I was twenty-three years old when I wrote my first blog post for her, and I had never really talked about my feelings surrounding donor conception with anyone. Actually, Elizabeth was the first person to ever ask me how I felt about being donor conceived. It had always been a taboo subject, both in my family and with my circle of friends. Blogging through FamilyScholars.org unleashed something in me that was both startling and therapeutic. It was through exposure to new ideas—and a great variety of ideas—that I was able to find my personal truth and dissolve my former feelings of isolation and confusion.

That said, speaking to journalists and reporters for the first time made me realize what a true risk it can be for people to expose themselves and their personal stories in a public way. Infertility, sex, family, and relationships are all very personal and private matters. But as we know, the personal is political. And what choices and actions adults make regarding sex, science, and reproduction will reverberate throughout their lives and their children's lives. We create and enforce laws, taboos, expectations, and reward systems regarding gender relations, personal autonomy, and sexual and reproductive freedoms. Today, we are at a critical moment in directing the beast of reproductive-technology policy. We cannot do it without the honesty and participation of our storytellers—the people who are living and breathing the challenges of third-party reproduction.

I want to sincerely thank all of the people who have contributed to AnonymousUs.org since its beginning in January of 2011. You have already given me a valuable education that I cherish, and I know your stories will do the same for others.

To contribute your story to the Anonymous Us Project, please visit AnonymousUs.org. We always appreciate learning from you, and each contribution represents a tile in the massive mosaic we are crafting called The Big Picture. Your voice will be heard in future volumes of this project, but as always, your privacy will be protected.

Welcome to the community.

Alana S. Newman
Founder, Anonymous Us

1. What Is Sperm Donation?

Sperm donation is the act in which a man provides his sperm to a woman who is not his sexual partner for the expressed intent of impregnating her. A sperm donor is the biological father of every child produced through his provisions, but he is generally never intended to be the legal father or to have much, if any, social contact with his offspring.

How Is It Practiced?

Sperm donation today is a multibillion-dollar industry. In fact, the word donor is mostly incorrect, because the vast majority of sperm donors are paid. A more correct title would be "sperm vendor," but for purposes of common recognition, I will continue to call these men "sperm donors" in this book. Most sperm donors are paid not only for their semen but for the legal relinquishment of the children their semen is used to produce.

The sperm-bank industry arose from the technology of **artificial insemination** (AI), which is where semen is introduced into the female for the intention of fertilization, usually directed by a third-party practitioner, by means other than sexual intercourse.

AI began as a means of breeding cattle and pigs. The first recorded use of artificial insemination with a human donor was around 1884, when a Philadelphia medical-school professor impregnated his unconscious and unknowing patient with the sperm of a medical student who was considered the best looking in his class. The woman conceived a son, believing the child to be the shared biological offspring of herself and her husband. The husband, thrilled that their infertility problem, which had been caused by his sexually transmitted disease, had been "solved," held nothing against the doctor and kept the secret from his wife.[1]

What was considered a highly secretive and private transaction then has sprung a lucrative industry in the global sperm trade today.

In addition to artificial insemination, there is **natural insemination** (NI), where a man and woman have sexual intercourse with the expressed intent of the biological father impregnating the woman yet relinquishing all legal parental rights and responsibilities. The sexual act is initiated as a service rendered, with no expectation of romantic companionship.

Who Is Using It?

The primary buyers of sperm are heterosexual couples experiencing male-factor infertility, lesbian couples, and single moms by choice. Originally, only heterosexual couples were allowed access to donor sperm, but gradually, sperm banks began servicing lesbian couples and more recently, single mothers and surrogates for male homosexual couples.

How Much Does It Pay?

In the United States, sperm donors often receive $400–$1,000 per month for up to two deposits of sperm per week.

What Does the Law Say?

The law is inconsistent throughout the Western world regarding sperm donation. Different countries have different maximum numbers of donations per donor. In Sweden, for example, a donor can only fertilize the eggs of up to six couples—each couple is granted a sibling, for a total maximum offspring of twelve per donor. In the United States, the American Society for Reproductive Medicine's guidelines suggest a limit of twenty-five children per donor per population of 800,000.[2] That ratio reduces to a limit of one child per donor per population of 32,000. That translates to the following measures:

- 258 offspring per donor within New York City,
- 608 offspring per donor within New York state,
- 9,812 offspring per donor within the United States, and
- 219,676 offspring per donor internationally.

Note that these numbers are merely guidelines. In the United States, there is no legal limit to the number of offspring one sperm donor can produce.

Some countries have outlawed anonymous gamete donation completely. In the UK, the Human Fertilisation and Embryology Act came into force on August 1, 1991, creating a central repository for donor information. Over a decade later, on April 1, 2005, anonymous gamete donation ended entirely, giving every British donor-conceived person, upon reaching sixteen years old, a right to information (including identifiable information) about his or her biological parent(s).[3]

Not all men are permitted to donate sperm. The most controversial ban on sperm donation was the 2007 U.S. Food and Drug Administration's ban on "men who have had sex with another man in the preceding 5 years," citing a significantly increased risk for HIV and Hepatitis B transmission.[4] Directed sperm donation from a known and preferred source, however, is permitted, no matter the sexual orientation of the donor.

How Can I Find Out More?

If you want to discover how sperm donation is viewed in mainstream culture, there are several movies in which it is depicted, including the following:

- *Delivery Man* (2013)
- *The Kids Are All Right* (2010)
- *The Switch* (2010)
- *The Back-up Plan* (2010)
- *Miss Conception* (2008)
- *Vicky Donor* (2012, English and Hindi)
- *Starbuck* (2011, French)
- *The Babymakers* (2012)

You can also find further education online:

- The Donor Sibling Registry: www.donorsiblingregistry.com
- Donor Conception Network: www.dcnetwork.org
- Infertility Network: www.infertilitynetwork.org
- "Confessions of a Cryokid" (blog): www.cryokidconfessions.blogspot.com
- "Donated Generation" (blog): www.donatedgeneration.blogspot.com
- "The Donor Who Dared to Say Don't" (blog): www.thedonorwhodared.blogspot.com

Notes

1. See "Artificial Impregnation," letter to the editor, *Medical World*, April 1909: 163–164. Accessed as part of Stephanie Blessing, "More on Artificial Insemination," blog post, FamilyScholars.org, February 17, 2011, familyscholars.org/2011/02/17/4579.
2. See Jeffrey A. Kuller, William R. Meyer, Kyle D. Traynor, and Katherine E. Hartmann, "Disposition of Sperm Donors with Resultant Abnormal Pregnancies," *Human Reproduction* 16 (2001): 1553–1555, doi: 10.1093/humrep/16.8.1553, introduction, in which the authors reference recommendations by the American College of Obstetricians and Gynecologists. See also "ASRM and SART Exert Pressure to Hinder Legislation in Several States," *ASRM Bulletin*, Febrary 20, 2012, www.socrei.org/ASRM_and_SART_Exert_Pressure_to_Hinde_Legislation_in_Several_States. For responses from the third-party-reproduction community, see marilynn [protected identity], "Cum Again? What's the Limit of Offspring per Donor?" blog post, *Relative Interests*, March 12, 2012, http://relative-interests.blogspot.com/2012/03/cum-again-whats-limit-of-offspring-per.html; wendykr [protected identity], "ASRM Fights Record Keeping, Tracking, and Limiting Numbers of Offspring to One Donor," blog post, *Donor Sibling Registry*, March 31, 2012, https://www.donorsiblingregistry.com/blog/?p=374.
3. See "If You Were Conceived After 1 April 2005," *Human Fertilisation & Embryology Authority* (last updated October 1, 2012), http://www.hfea.gov.uk/5526.html.
4. U.S. Food and Drug Administration, "E. What Risk Factors or Conditions Do I Look for

When Screening a Donor?" under "IV. Donor Screening," *Guidance for Industry: Eligibility Determination for Donors of Human Cells, Tissues, and Cellular and Tissue-based Products*, August 2007, www.fda.gov/BiologicsBloodVaccines/GuidanceComplianceRegulatoryInformation/Guidances/Tissue/ucm073964.htm#DONORSCREENING1271.75.

2. What Is Egg Donation?

The first successful egg-donation procedure in the United States was executed at Harbor UCLA Medical Center in July 1983; the baby was born on February 3, 1984. The modern egg-donation process involves doctors removing viable eggs (oocytes) from a young woman usually in her twenties, fertilizing them into embryos outside of the body, and implanting the embryos into the womb of the recipient female (either a birthmother or surrogate). Also, sometimes eggs are given to create embryonic stem cells for biomedical research.

The egg donor is the genetic mother of the child, but the woman who gives birth to the baby is the birthmother and is also considered a biological mother, unless she's for hire, in which case her title is changed to surrogate or gestational carrier. If the child born is discovered to carry any genetic diseases, it will be the egg-donor mother's medical history that will be most helpful to the offspring. In so far as a child physically resembles his parents, the child will more likely resemble the woman who provided the egg for his conception.

How Is It Practiced?

In America, an egg donor must undergo physical and psychological evaluations to determine her eligibility. One of the core concerns is that she'll show reliable compliance and be able to complete the physically and emotionally difficult process of egg harvesting. Once she passes the screening, she begins the three-to-six-week process of synching her menstrual cycle with that of the birth mother, by taking oral contraceptives. Once the cycles are synched, the egg donor self-administers a series of hormone injections daily for one to three weeks, stopping the normal function of her ovaries. Then she begins self-administering a series of egg-follicle-stimulating hormone injections, which will dramatically increase the number of eggs she produces in preparation for the harvesting. Roughly thirty-six hours before retrieval, the egg donor will inject the hCG hormone, which works to mature the eggs. The egg donor must be sedated during retrieval, which takes twenty to thirty minutes. The doctor inserts a probe through the vagina that suctions out each egg follicle.

The bodily stress and potential risks associated with egg donation are extensive. Increasing numbers of women are dying or suffering critical injury from complications of the procedure. Shockingly, there have been no long-term studies on the health outcomes of egg donors,[1] but these women, and their surviving families, are beginning to speak up.

The biggest risks (that we know of) include OHSS (ovarian hyperstimulation syndrome), which in severe cases is lethal; increased potential for developing ovarian cancer; and future infertility.

Who Is Using It?

Egg donation is associated with roughly 5 percent of all in-vitro-fertilization (IVF) procedures (see ch. 4). Heterosexual couples that use egg donation usually do so because of age-related infertility, to avoid passing down any genetic diseases, or because of complications from other medical conditions like cancer. Gay and single men are one of the fastest-growing markets for female eggs because of the decreasing stigma of gay- and single-fatherhood scenarios.

Most egg donors are anonymous and young, aged nineteen through twenty-nine. Marketing efforts are largely focused on college students who carry student-loan debt.

How Much Does It Pay?

Laws surrounding payment of egg donors vary across the world. In the UK, they may only receive up to £750 (about $1,150) to cover "costs incurred." In Sweden, payment for costs incurred is roughly $1,500. But in the United States, the average amount of compensation for an egg "donor" is $8,000–$9,500 per procedure.[2]

Human eggs have become the most valuable commodity on the planet by weight and volume. Although U.S. women are not paid on a "per egg" basis, the average payment they receive divided by the average number of eggs they produce equals roughly $700 per egg.

In order to circumvent low supplies of eggs due to regulation and the relatively small amount of monetary compensation for British egg donors, some UK clinics offer "egg sharing," where a woman who wants to undergo IVF in order to conceive her own children will pay less, little, or nothing in exchange for donating a number of her eggs or embryos, either for research or human reproduction.

What Does the Law Say?

Egg-donation laws are different from state to state and country to country. Some places have banned donor anonymity, although parents are not specifically required to tell their donor-conceived children the truth about their conception. All birth mothers are automatic legal mothers in nearly every state and Western country, but sometimes, an egg donor's citizenship status can affect the citizenship rights of the child. Most authorities suggest the use of a reproductive lawyer when purchasing and utilizing donor eggs because of the legal complications from region to region.

How Can I Find Out More?

If you wish to further explore this topic, visit eggsploitation.com to learn about "Eggsploitation," a documentary that won the California Independent Film Festival. Another website, www.WeAreEggDonors.com, was founded by a team of egg donors and is dedicated to neutrality on the topic.

Notes

1. See Serena Gordon, "Risks and Benefits of Egg Donation Reported," *U.S. News & World Report*, December 26, 2008, http://health.usnews.com/health-news/family-health/womens-health/articles/2008/12/26/risks-and-benefits-of-egg-donation-reported.
2. See Aaron D. Levine, "Self-regulation, Compensation, and the Ethical Recruitment of Oocyte Donors," *Hastings Center Report* 40, no. 2 (2010): 25–36.

3. What Is Surrogacy?

Surrogacy is when a woman is impregnated and then carries and delivers a baby with the purpose of granting all parental rights to another couple or person. With **traditional surrogacy**, a woman is impregnated either naturally or through artificial insemination (see p. 13) and delivers her own genetic child. With **gestational surrogacy**, a woman is impregnated using **in vitro fertilization** (IVF) (see ch. 4); the egg belongs to another woman, and the baby the gestational surrogate delivers is not genetically hers.

Most surrogacy arrangements today are gestational arrangements, because traditional surrogate arrangements are considered emotionally and ethically complicated. There have been many instances where traditional surrogates have found it difficult to give up their biological child.[1] Many consider traditional surrogacy as baby selling, which is a form of human trafficking and supposedly illegal.

How Is It Practiced?

India is by far the most talked about and popular location for fertility tourists interested in surrogates. Indian surrogates are paid far less than those in the United States, but it is still roughly ten years' salary for rural Indians.

Indian women typically keep their identity as surrogates secret from friends and relatives and create false alibis to veil their secret venture. During the duration of their pregnancy, they might live in dormitories with ten to twenty women sharing one room. Their diet, exercise, and health supplements are strictly monitored and regulated, and most deliveries are through scheduled Cesarean section, allowing commissioning parents to schedule their travel arrangements succinctly with the arrival of the baby.

In the United States, the largest demographic of surrogate mothers are military wives. They have generous healthcare coverage paid for by American taxes, and their husbands are often on duty abroad and thus sacrifice little in terms of sexual access to their wives.

Surrogate mothers' families are usually near complete—meaning they have had all the children they intend to have. And they are typically low income and motivated by the payment.

Who Is Using It?

The intended parent(s) may commission a surrogacy pregnancy for a variety of reasons. Male homosexuality, female infertility, or lack of a partner (single fathers by choice) are all circumstances that motivate use of a surrogate. Zippi Brand Frank, director of *Google Baby* (2009), predicts that in the future, "surrogacy will not only be used as a last resort but rather as an alternative for women who do not wish to have stretch marks or might not want to be pregnant because of their careers."[2]

How Much Does It Pay?

In some countries and states, it is illegal to pay surrogates, and they must only do it altruistically. In the United States, the typical surrogate fee is between $40,000 and $70,000. In India, her fee would be $5,000 to $12,000.

What Does the Law Say?

Surrogacy is illegal in many states and countries, but it is most commonly practiced in India and the United States. It is essential for intending parents to hire a reproductive law/surrogacy attorney, due to the complex legal/ethical landscape and global nature of surrogacy. Variations on legality include legality of payment, citizenship issues for the child, what to do if a fetus is abnormal or threatens the life of the surrogate, what to do if the intended parents split up before the child is born, and countless other considerations.

In early 2013, India instituted significant regulations regarding surrogacy, making it only available to heterosexual couples who have been married for at least two years. Couples are also required to apply for a medical visa, whereas in the past, tourist visas sufficed. Further, the new set of laws demand that the recipient parents be citizens of a country where surrogacy is legal and recognized—to avoid "stateless" offspring.

How Can I Find Out More?

Three documentaries that explore the lives of surrogates are *Frozen Angels* (2005), *Google Baby* (2009), and *Made in India* (2010). These are all either in English or

available with subtitles.

Notes

1. For one early, dramatic example, see "Baby M," *Wikipedia*, en.wikipedia.org/wiki/Baby_M.
2. Zippi Brand Frank, "Director's Statement," www.zippibrandfrank.com (click on "Google Baby 2009").

4. What Is IVF?

In vitro fertilization (IVF) is the process by which an egg is fertilized by sperm, creating one or several embryos, outside of the body. "In vitro" is Latin for "in glass."

With IVF, a doctor removes a woman's egg(s) from her ovaries and then fertilizes the egg(s) in a fluid container with retrieved semen. The doctor then implants the new embryo(s) into a woman's womb, usually that of the embryo's genetic mother. It is the IVF process that has made gestational surrogacy possible and spawned a booming industry in surrogacy and egg-donor arrangements.

The process was invented by Dr. Robert G. Edwards, a British physiologist who won the Nobel Prize in Physiology or Medicine in 2010 specifically for his advancements in IVF. In partnership with Dr. Patrick Steptoe, a gynecologic surgeon, he engineered the successful birth of Louise Brown, the first IVF baby, in 1978.

How Is It Practiced?

IVF requires the use of strong fertility drugs to stimulate egg-follicle maturation (these are the same hormones used in egg donation). There are medical risks involved for the woman undergoing treatment, but because she is usually both the genetic mother and the legal mother of the future child and will receive any and all benefits should the procedure be successful, the ethical dilemma is greatly simplified.

The year 2012 marked over five million births via IVF treatment.

IVF often entails use of donor egg or sperm, but it is typically performed with parental egg and sperm. People conceived via sperm donation are often mislabeled as "IVF babies" or "test-tube babies" when in fact, sperm donation usually involves insemination rather than in vitro fertilization. Offspring conceived from egg donors, however, are conceived using IVF, by definition.

Sometimes errors do occur due to the microscopic nature of sperm and eggs. There have been a number of births where the "wrong" sperm and eggs have been mixed, accidentally resulting in a child genetically unrelated to one or both legal parents.

During IVF, there are often embryos created that are not "used" for whatever reason. There are roughly half a million embryos frozen in storage today in America that if implanted would potentially grow into fully independent people.

Who Is Using It?

IVF has allowed millions of women to experience pregnancy and give birth to their own genetic children. It has also allowed for thousands of infertile women, gay men, and single men (via surrogates) to procure children, genetically related or not.

How Much Does It Pay?

IVF is a laboratory science that requires extensive training and access to precision medical instruments. Hopeful parents pay doctors and facilities just like with any other elective medical procedure, and costs vary.

What Does the Law Say?

One country in the Western Hemisphere with a history of banning IVF is Costa Rica. But the Inter-American Court of Human Rights struck down the ban in December 2012, stating that they are violating the human rights of infertile couples by not allowing them IVF treatment.[1]

Most U.S. laws focus less on regulation and more on antidiscrimination provisions that require doctors to perform treatment if, for instance, they have objections to using IVF to create children for single parents by choice or same-sex couples.

How Can I Find Out More?

Robert G. Edwards and Patrick Steptoe wrote their own book describing the long road to engineering the IVF procedure and the birth of Louise Brown: *A Matter of Life: The Story of IVF—a Medical Breakthrough.*

Notes

1. See Associated Press, "Costa Rica: Ban on In Vitro Fertilization Is Struck Down," *New York Times*, December 21, 2012, www.nytimes.com/2012/12/22/world/americas/costa-rican-ban-on-in-vitro-fertilization-is-struck-down.html?_r=0.

5. What Is Adoption?

Adoption is an institution where social workers and agents vowing to work in the best interests of children secure stable homes for those whose natural parents are unable or unwilling to care for them for whatever reason.

There are two main types of adoptions: **Open adoptions** allow for contact and communication, or at least the exchange of identifying information, between adoptive families, adoptees, and biological families. Open adoptions maintain an adoptee's rights to their unaltered birth certificates and/or adoption records. **Closed adoptions** seal all identifying information; the identities of adoptive families and biological families are kept secret. Some adoptions employ a combination of these practices, such as when communication is moderated by a third party or an adoptee is granted access to records at age eighteen.

How Is It Practiced?

During antiquity, true adoption of infants was rare; adoption was typically practiced for adults as a way of transferring wealth. Abandoned children were mostly used as slaves. In the Middle Ages, European culture was anti-adoption. Bloodlines were paramount, and biologically unrelated "foundlings" were often brought in as servants. Churches were the first to care for abandoned children in any institutionalized way. The first modern adoption laws in the United States where the best interests of the child were noted were enacted by the Commonwealth of Massachusetts in 1851.

Modern adoption in America began as a solution to the unprecedented number of orphans and abandoned children following rapid nineteenth-century immigration as well as the Civil War. Touted as "the best method of disposing of our pauper and vagrant children," the Orphan Train Movement transported 200,000–250,000 children from East Coast cities to rural America in what became the largest migration of children in history. Many of the children were not raised as family members but servants. Because of the magnificent size of the movement and the high level of exploitation involved, new agencies and laws emerged promoting adoption more as we know it today, where children are raised within nuclear families as if they were genetically related members.

The period after World War II marked a loosening of sexual mores. Subsequently, nonmarital births tripled in number. At the same time, the scientific and political community began stressing the idea that nurture determined more about character and personality than did nature. This was in response to the troubling Third Reich emphasis on Aryan superiority and the "perfecting" of humanity. The new emphasis on nurture served to increase positive views on adoption, elbowing out previous anti-adoption eugenic ideology, which assumed orphans and nonmaritally conceived people to be inferior.

The number of adoptions in the United States peaked in 1970 and has been in rapid decline since then. This is likely due to the introduction of birth control pills, the legalization of abortion, a steep decline in stigma for illegitimate births and unwed mothers, and the proliferation of artificial reproductive technologies.

Who Is Using It?

Adoption is most used by infertile couples, and women often lead the pursuit. Some couples, often religious, desire adoption even if they are fertile, because they find value in providing for and nurturing a child whose genetic parents were unable to do so.

How Much Does It Pay?

It is illegal in the United States and most Western countries to pay a birthmother to relinquish her child. However, adoptive parents typically do pay for costs associated with the pregnancy and delivery as well as home-study fees, travel expenses, agency fees, etc. Adoption through foster care is the least expensive route to parenthood, and foster parents may even receive monetary stipends and tax credits for finalizing an adoption and providing a child a permanent home. It is also generally less expensive to adopt children considered "harder to place," such as older children, children with special needs, or children of minority race. Domestic adoptions are significantly less expensive than international adoptions.

What Does the Law Say?

There are four parties that regulate adoption: the government, adoption agencies, birthmothers, and adoptive parents.

Besides certain legal paperwork, **the U.S. government** requires home studies of all potential adoptive parents. This is the essential test one must pass. It evaluates the home, lifestyle, and parenting abilities of the adopter to ensure safety and well-being for the child.

Adoption agencies reserve more subjective rights concerning whom they allow to adopt through their services. Most agree that an adopter should be at least eighteen years old in order to properly care for a child, but there are also maximum age requirements, typically capping new adopting parents at forty years old. This is to ensure that the child will have living, healthy parents until she reaches adulthood—especially through the troublesome teen years, when identity struggles may become extreme and the adoptee is in most need of special support. Adoption agencies are also allowed to restrict their services based on marital status and religious affiliation.

A **birthmother** is allowed to impose highly subjective criteria on adopting parents before relinquishing her parental rights. She is allowed to discriminate on her own terms and might get to know adoptive couples intimately before agreeing to or refusing to terminate custody of her offspring.

The birthfather's rights are typically secondary to those of the birthmother. Because he experiences his fatherhood outside of his body and the birthmother experiences her motherhood within her body, she will have much more practical control over what happens to the baby. She may not even choose to tell the father about the pregnancy.

Adoptive parents are not required to adopt any child or agree to any variable set of circumstances. They have control and veto power up until the point the custody paperwork is signed, after which they are fully responsible for the child they adopt. They may agree to certain conditions, but once the adoption is complete, they are the legal parents and hold all the cards.

It is illegal to coerce a birthparent into relinquishment with money or threats.

It is illegal to traffic children or to intentionally impregnate a woman for the purpose of adopting out the child.

6. Themes from Donor-conceived Persons

Censorship

When some donor-conceived people try to speak out in criticism of the practice of third-party reproduction (3PR), we are met with words like, "But would you rather not exist?"

Being donor-conceived can seem like an existential gag order—as if becoming an embryo and eventual adult human was a nonverbal contract of approval for the means by which we were conceived. Many of us are swept off our platforms by a kind of moral relativism; because other people "have it worse," we're not allowed to complain. Because "everybody has some issues in life," we're not allowed to voice ours. Many 3PR proponents push a position that all life is a blessing and that we should absolutely be thankful for ours. But is the moral wrong of an act neutralized if something beautiful is created from it?

Arguing against donor conception as a donor-conceived person often means arguing against your very existence, which is an exercise in advanced philosophy. I believe we should begin framing ethical debates regarding 3PR in similar fashion to other non-ideal forms of conception. For example, we would never ask people conceived via rape or incest why they're ungrateful for life when and if they critique the methods of their conception.

But by far, the strongest source of censorship is the pressure toward secrecy within our own families:

> [My mother] makes me feel as though I should not have any feelings on the subject of being donor conceived.... She never will understand how I feel. I wish she would understand. I don't want her to blame herself for my misery. I try not to broach the subject at all with my nonbiological father.... I am a young woman who is asked to keep quiet about her innermost conflict ("Alone," p. 85).

Many of us feel we must honor our family's pride and withhold from airing the dirty laundry. Criticizing donor conception itself is hard to separate from directly criticizing our parents:

> I feel silenced for so many reasons: because it's weird and people don't get it, because I don't want to hurt my dad—whom I absolutely adore—talking about it, and also because I am always afraid I will burst into tears talking about it, no matter who is listening. I also know that my parents are very private people in general, and to have a bunch of people—whoever they may be—know that I am donor conceived would make them uncomfortable ("Silence," p. 93).

And unfortunately for some, disobeying or dishonoring custodial parents can threaten survival itself—where support, assets, and contact are dangled in escrow as incentives for compliance in the lie that everything is fine and normal.

Inconsistency in Law

Not only do different countries and states have different laws on a range of important issues regarding donor conception, but those laws may even contradict other laws within a country, leaving donor-conceived people feeling cheated and discriminated against. For example, in some places, a child conceived after a given date can access their genetic parent's identifying information—but not children born before then. In some countries, anonymity is banned altogether, while in other places, like California, it is the encouraged norm. The United Nations Convention on the Rights of the Child states, as summarized by Global Governance Watch, that every child has the right to "his or her own name and identity, to be raised by his or her parents within a family or cultural grouping, and to have a relationship with both parents, even if they are separated."[1] This policy seems to contradict third-party reproduction entirely. There is also an optional protocol (secondary extension) of the UN Convention on the Rights of the Child that prohibits the sale of children. The indisputable authority of these statutes has required some 3PR proponents to perform verbal gymnastics to justify the use of third-party reproduction.

Many donor-conceived people voice concern over the conflict between laws regarding donor conception and those concerning adoption or other child-welfare issues. For example, adoptees are guaranteed a certain standard of care by the home-study requirement, which screens nonbiological adoptive parents. Donor-conceived people do not have this safety measure and thus are at greater risk for being raised by unfit social parents. Also, children conceived traditionally are guaranteed a right to child support if one of their natural parents rejects custodial duties. Donor-conceived children do not enjoy the same assurance.

For many donor-conceived people, these laws and statutes demonstrate that we are being discriminated against or, at best, politically and legally neglected.

The Cinderella Effect and Low-quality Parenting

Many storytellers praise the quality of care they received growing up and showcase a remarkable amount of love toward their custodial parents. However, there have also been submissions relaying instances of abuse—where the basic safety of the child was in jeopardy. The Cinderella Effect is a well-documented phenomenon that tells us that statistically, nowhere is a child safer than with his biological/genetic parents. Abuse and negligence rise dramatically when children are in the care of one or more nonbiological guardians—the default circumstance for many a donor-conceived child:

> [H]e started touching me inappropriately when I reached out to horseplay. When I told him to stop, he said I didn't know what I was talking about, and we were just having fun. I stopped horseplaying. I stopped doing gymnastics around the house. I stopped feeling free in my body. I stopped being a kid ("Safety in Biology," p. 87).

More surprisingly, we must ask if children are even safe with the biological parent who procured the sperm or eggs. One contributor shares the story of her first memory—that of her own biological mother trying to strangle her in her crib:

> If my grandmother had not arrived unexpectedly that afternoon and interrupted my mother, I would have become a cot death statistic, but nobody will want to believe that my own mother had tried to kill me, especially as she had spent five years and a lot of money desperately trying to have me ("Beginnings and Ends," p. 86).

Is the desire to have a child really all we require when allowing anyone to create a life through reproductive technologies? Is this justifiable, given that all adoptive parents must undergo home studies and screening before they receive a child? Is it possible that some individuals (like "Octomom" Nadya Suleman) may be putting the children they create at great risk? Without the market-like industries that are fertility clinics and sperm banks, would these commissioning parties have been approved as parents, either by the ruthlessness of the dating scene or the reasoning skills of an adoption agent? Is it possible we are doing a disservice to the coming generation by failing to regulate this industry and screen intending parents?

The Great Dad

Of the stories I've heard, both in and out of this book, I am convinced that great social fathers (and mothers, together as a team) are the most powerful tool in mitigating the negative threats of donor conception. While I know donor-conceived people who had great dads and are still troubled by 3PR and where it's going, clearly the involved and loving social parents are the most important factor in making happy, well-adjusted kids:

> As others have stated feelings of abandonment by their donor, I do not feel this way. I actually have no interest in meeting this man or forming a relationship with him....I have my daddy, who has always been there for me. I have a dad, and he is irreplaceable, although he has his issues and may not have been an intelligent doctor. He has been the best dad I could ask for....He always treated me as if I were his own blood and called me his daughter and been proud of my accomplishments like any other dad would be. For me this has worked in a different way than for most. Finding this out now made me realize how much I love my daddy, and I don't care if I'm not genetically his ("I Love My Daddy," p. 71).

The Ill Dad

The good dad in donor conception is definitely the unsung hero. But many of these dads were forced to use donors because of sickness, and I hear many stories from donor-conceived persons about their social father's complicated health situations or early death, and the emotional complications of only being told about their conception beside their dad's deathbed: "I grew up in constant fear of many diseases which run on my dad's side of the family" ("Completely Shocked," p. 77).

We should be asking, What is going on with our bodies? Why are we so sick? Infertility is when environmental and medical circumstances collide to prohibit the body from healthy, natural reproduction. Something is wrong when 5.3 million people are infertile and sperm count has dropped by as much as 50 percent in the last fifty years.[2] Have we so totally disembodied ourselves to the point we forget that we live within our ecosystem and that we ourselves absorb all the toxins that we pump out? One donor-conceived person identifies pollution as the antagonist of his family woes. Pollution made his social father infertile; therefore, pollution is to blame for his parents using donor sperm. If we act to improve our environment, are we also acting to curb infertility?

> My "social father" had a rare disorder where he had three chromosomes: XXY. This means he was basically male and female. He had a male body, but during puberty, the estrogen secreted by his female-informed endocrine system basically rendered him sterile....Essentially, if you're a mammal and you have a metabolism, you have been exposed to endocrine-disrupting toxicity, and your immune system, physical reproductive capacities, and social/motivational reproductive capacities have been compromised ("Endocrine Disruptors," p. 73).

Common Language

There are many common phrases and remarks I've picked up from both the submitted stories and in meeting donor-conceived people in real life. Many of us, upon being questioned about our fathers, are inspired to lie and tell people he died. This usually works to deflect further questioning, making people uncomfortable—though significantly less uncomfortable than if we were to tell them the truth.

Many of us describe our missing biological fathers as half of whom we are. We use words like "holes," "missing," "incomplete," and "void" in describing how being donor-conceived has affected us.

Donor-conceived people express varying degrees of support or disdain for the practice of gamete donation, but even those who are supportive will say things like, "I just wish I had a picture."

Money a Low Priority but a Big Issue

For sperm donors out there concerned that their offspring will come knocking on their door begging for money, you need not fear—there is ample anecdotal evidence that we have no feelings of entitlement to your cash. Many donor-conceived people go out of their way to vocalize that they are not interested in money:

> I don't want back child-support payments, as some donors fear. I don't want his love or to call him "Dad"—I already have a dad. I don't want to be on his family's Christmas cards or to take up an inordinate amount of his time. I just want to know who he is ("How My Mother Told Me," p. 80).

But we are troubled by the exchange of money in our conception. To many of us, donor payment reeks of baby selling, prostitution, and human trafficking. It's much harder to accept yourself and to respect your parents (both genetic and social) when money is involved in your conception:

> Screwed on the lid
> Of the lives I'll never know
> Golden handshake
> Of the human industry ("The Cup," p. 65).

Another writer asks his ethereal father, "Do you honestly care about the poor couples out there unable to conceive? Or did you just need the money?" ("A Letter to My Father" p. 62).

Moving On

Many donor-conceived people reject any kind of victim status and urge their troubled peers to move on with their lives and find meaning and purpose for their existence. This is a legitimate and practical perspective that has its merits:

> I have never met my donor "father," and I have no desire to do so. I do not see this lack of contact with my biological father as something missing in my life, and I have no hurt at the fact of my creation. What does

> cause me hurt, however, is the idea, constantly repeated by small numbers of donor-conceived children, and in popular media representations of the issue, that there must be something wrong with your life if you do not know your biological father....Knowing your biological parent will never explain why you love the one you do, why you hate early mornings or feel uplifted by classical music. Genes interact with your environment to create you, but they do not give meaning to your life ("I Have Never Met My Donor 'Father,' " p. 66).

Despite the above contributor's claim that only "small numbers of donor-conceived children" feel that something is missing in their lives, the fact is that until 3PR is properly documented and accurate studies are conducted ("My Daddy's Name Is Donor" was the largest study of its kind but was criticized for not being peer-reviewed), we won't know what percentage of offspring are well adjusted and at peace with their conception.

Notes

1. "The United Nations Convention on the Rights of the Child," *Global Governance Watch* (undated), www.globalgovernancewatch.org/human_rights/the-united-nations-convention-on-the-rights-of-the-child.
2. See Theo Colborn, Dianne Dumanoski, and John Peterson Myers, *Our Stolen Future: Are We Threatening Our Fertility, Intelligence, and Survival? A Scientific Detective Story* (New York: Plume, 1997).

7. Themes from Adoptees

I chose to invite stories from adoptees because I so often hear things like, "Donor conception is just like adoption, isn't it?" Both are means by which people desiring children act to procure them, and both inherently mean the separation from a biological parent. If adoption is okay, then donor conception must be okay too, right?

Adoption as we practice it today is an institutional response to the tragedy of orphans, child abandonment, and failed parenting. Adoption does not primarily act as a *market* to provide children for adults who want them—although many adoptees'-rights organizations argue that current systems are abusive and market-like. But when practiced ethically, adoption is supposed to find the best possible parents for *existing* children who desperately need a loving home.

Most adoptees have had the privilege of being put into families who have passed rigorous home studies; we can expect that they experience a reliable degree of safety and support from their adoptive parents. Is there a noticeable difference when separation from biological parents is caused by human frailty and unintended circumstances vs. deliberate design where medical entrepreneurs and lawyers actually *profit from* and *promote* the separation (as with donor conception)? This is a question I would like to see addressed.

My personal experiences growing up with an adopted sister tell me that these things are complicated: while some people will never go out of their way to find out more about their birthparents, others would travel to the ends of the earth. I want to thank those adoptees who shared pieces of their stories with us through Anonymous Us and invite more of you to chime in so we can better know you and understand what life is like for you. These are some of the themes and ideas I heard from our contributors:

Longing to Know

Loving, stable homes appreciated and noted, many adoptees still are curious about their biological parents, affirming the natural desire to know where one comes from: "Curiosity mixed with longing becomes an itch, and the more I scratched, the more I searched, the more I uncovered a wound. Shut out, locked out, told I had no right to know, I railed against the walls that had been erected around my identity" ("The Missing," p. 123).

Searching for Truth

As sentient beings, we know that the universe does not suddenly begin when we're born or when we begin remembering. We know that what we can expect and prepare for in the future may largely be determined by the past. Adoptees very often desire to know the truth about their family history, starting from before their adoption—even if that truth is uncomfortable: "The truth is something I had to know for myself. In fact, the truth was much more interesting than anything I imagined. It was the dreams and fantasies that got to me. What right does an adopted person have to the truth? The same as everyone" ("It Was About Truth," p. 122).

Without truth and access to our stories, access to our own grief, people are left to live in a false reality, punishing themselves when they struggle to live up to fantasies that everything is okay now that the parents have a child and the child has parents: "Trying to conform and fit like others, I was shaped like clay by adoption's fingers—by secrets. Silences pressed, creased, accented. So much in private, hidden from me. I know things about solitude. I'm still trying to find words" ("Isolation," p. 121).

Disenfranchised Grief

Grief is disenfranchised when society does not acknowledge it. In response to someone vocalizing personal grief, a disenfranchising response could be "Get over it. Move on. You can't dwell on it forever." It's difficult for society to acknowledge all forms of grief, because to do so would mean having to be willing to then do something to either assuage or relieve it—or to prevent further grievances from happening in the future:

> [M]ost of society *doesn't want* the *real me*. They want the "Happy to be Adopted" me. The one who, sure, may have a *hint* of *curiosity*, but nothing that really affects me deeply—because I am fulfilling the role that was written out on my amended birth certificate, my adoption decree—that is *who I am supposed to be* ("Finding Me in the Bathroom," p. 124).

Forgiveness and Acceptance

Of course, forgiveness and acceptance are practical and helpful suggestions for anyone encountering grief. Resentment and torment isn't healthy for the carrier. Pain happens, and we must eventually move on with our lives to find peace and purpose. But one contributor considered some of her genetically alienated peers to be the authors of their own grief. She advised them, "Find your strength from inside. Adoptees/donees who were raised by loving parents only have a 'hole' or a 'missing piece' if they decide to create that hole themselves" ("Don't Drive Yourself Crazy," p. 122).

Pain can be especially hard to deal with when the source of it all is your protector, your *parent* whom you love and feel loyalty toward. If we want to continue to have a healthy relationship with our functional families, we must learn to forgive them for their imperfect actions and judgment:

> I now realize that my mother wanted me and planned for me, just not in the best way. We are working on our relationship, and I have since become a Christian, and through the Lord I am healing....I am now thirty-three and beginning to learn to love my mother and forgive those who have hurt me. We have a long way to go, but I can accept that I am loved and was not a mistake ("Desperate for a Child," p. 119).

I'd like to add that building Anonymous Us has been very therapeutic for me. I also wrote a screenplay where I was able to construct a more complete sense of my biological father through narrative storytelling. Your imagination can be a powerful tool for good, and I encourage everybody who is having a hard time letting go of *any* kind of pain, to embrace it and *build*—it's the fastest route to redemption.

8. Themes from Donors and Surrogates

Who are these people—these mysterious donors and surrogates who agree to create new life by contract? Can I be proud to share blood with them? Can I be proud to have been one? Why did they do this? How did they frame it for themselves? What is their worldview?

I actually was an egg donor myself. It was at a time in my life when I was opposed to anonymity but was still operating under the script I grew up with—that 3PR was a normal and ethical way of making babies. I sold my eggs as an open-ID donor, and according to the "My Daddy's Name Is Donor" report, donor-conceived people are twenty times more likely to go on to sell or donate their own gametes as adults, so I'm not alone.

Since my "donation," I have yet to receive any follow-up from the clinic or egg-brokering agency asking how my mental or physical health is doing. I'd like to give donors and surrogates an opportunity here to tell us how they feel and how we can better serve their interests.

"Like" a Parent?

Albeit sperm donors admit they are absent, they still use words like "father" and "my child" to describe their relationship to their biological children. Egg donors seem to reject any notion that they are the child's mother, even though they have equal genetic relatedness to their donor children, and even though the act of retrieving eggs is substantially more time consuming and riskier to one's health than procuring sperm. One egg donor writes, "The children who have my genes are not my children. They are my genes. Their mother is their mother" ("So Wonderful!" p. 109).

However, this is different from an adopting mom, or a mom who used a surrogate, who might say that the woman who gave birth is *not* the mother. A woman who used a surrogate to carry her own genetic child after IVF might even claim that it is the fact that *her* own egg was used that supports her claim to motherhood. Consequently, we are cross wiring terminology on a grand scale. People conceived using different forms of donor gametes and IVF or surrogacy may easily get confused about who is a member of our family and who is not.

And the situation can be just as confusing and frustrating for *users* of reproductive technology. For instance, I once met a lesbian who fertilized her own eggs using IVF with the sperm from a known donor and then had the embryo implanted into her romantic partner's womb. She complained at the ridiculousness of having to adopt her own genetic child, disappointed that she could not list both her partner *and* herself as mothers on the birth certificate.

The cases get more and more confusing. Melanie Thernstrom used an egg donor and two surrogates to bring forth her adorable "twiblings." The surrogates then later helped by breastfeeding the babies. Still she considers herself "the only mother" and that "actually, there is no biological mother."[1]

When society is this fractured and confused about roles and titles—who is important, who is a parent, and who is not—can you imagine how the kids feel?

> In their eyes you find my eyes, in their smiles, in their very body and even in their psyche, I am there as half of their heritage. This makes me feel that I live in them forward, my very core of being has been passed on.... I genuinely wish for my conceived children the best possible future ("I Am a Sperm Donor," p. 102).

It seems likely that sperm donors still recognize themselves as their children's fathers because for millennia, men have experienced parenthood outside of their bodies. Countless men have impregnated women and then abandoned ship, never getting to look into their child's eyes. Women, on the other hand, have *always* had to endure pregnancy and labor in order to reproduce. This new 3PR reality disrupts our framework. The egg-donor genetic mother is actually more like a father than ever before. For the first time in human history, women are reproducing *outside* of our bodies.

Motherhood is getting chopped up, tangled, and dismembered. When every human society on earth has a unique word for mother and for father, does it dehumanize us and disconnect us from our humanity when we scramble to replace, omit, and redefine such important words? Or are new words, and reproductive teams of people, all with different titles, a benign and welcome progression?

Lying

More than one contributor has admitted to lying in order to make it past the filtering process to become a donor. This comes as no surprise of course; just as people lie in order to get someone to sleep with them, donors may lie to get someone to buy their gametes. There's a lot of money (and pride) on the line.

> I realized a majority of the donors in some of these online forums were lying about their education, height, how many they fathered, type of employment, etc.…I started out lying just like some of these guys. Guess the only thing I was honest about was the way I look and that I'm STD free ("It All Started Out with Lies," p. 110).

Part-time Work

Many donors and surrogates are looking for work that pays well but gives them a lot of freedom. Many contributors speak of other obligations or of needing a way to pay for everyday expenditures without sacrificing a lot of time. Female contributors speak of wanting to stay at home and spend time with their kids but needing a way to pay the bills. Others recall selling their gametes to pay for luxury goods like clothing, entertainment, and status items. But it is most common to justify selling one's gametes to pay for higher education.

> I worked retail and waited tables and I still couldn't keep up with my peers. Rent, clothing, entertainment, travel expenses, cell phone, car costs, and student loans just kept adding up. I went on Craigslist and started looking for extra work.…When I "donated" my eggs, I did it because I needed the money ("Relief from the Real Problems, Followed by Regret," p. 110).

The Gay Sperm Donor

Many men without prospects for starting a family the traditional way go on to donate their sperm, such as gay men. They lament not being able to have a regular family due to their lack of sexual attraction to women, yet they still yearn to have children, or at least pass on their genes. They view sperm donation as an opportunity to have children, albeit children they will likely never know personally: "I am a gay man. And being gay, having accepted myself as such in the past three years, I have absolutely not come to accept being possibly robbed of the biggest gift of life, by design: a family, particularly made up with my own produced children" ("I Am a Sperm Donor," p. 102).

Some donors point out a lack of financial resources:

> I was in my midthirties, realizing that I might never have children of my own. It all started with me just browsing the search engines and came across some sperm-donor forums ads.…As for why I'm a sperm donor—basically I couldn't afford to have children of my own. I work a low-paying job like so many people ("It All Started Out with Lies," p. 110).

Because of the desire for fatherhood and the decrease in stigmatization for LGBT parenthood, gay men are now converting from being providers of sperm to being purchasers of eggs and surrogate wombs.

Joy from Donating

A thin majority of donors and surrogates described positive feelings regarding their contributions. There were words of joy, pride, happiness, and excitement. Many donors believe that a person or couple's eagerness to parent translates to an elevated capacity to parent. They are confident the child will be loved and safe and well taken care of: "I felt that if a single woman or a couple were prepared to go to the lengths of being treated with donor sperm, then the resulting child would be wanted and loved" ("Proud to Have Donated," p. 97).

> Donating my eggs was one of the most satisfying things I have done in my life. Knowing that I gave life to someone who otherwise was unable, brings me great happiness....I am proud of my choices and would be interested in meeting these children at some point, given that they and their parents would like to meet me ("So Wonderful!" p. 109).

Guilt from Donating

Other donors had a harder time, showing that the emotional journey of being a donor is fluid and can transform over time: "I feel grief. I am mourning for my lost two children, whom I never will see, never know more about them than the fact that they came to exist. I am mourning my role in the process, that it has just ended where others' is just starting" ("Yesterday I Got a Phone Call from the Fertility Institute," p. 104). Some donors felt shame: "I have a young child out there now that I can't see or know because of my selfishness and my inability to feel 'good enough' without a degree, without the clothes, without the car....I'm sorry to the child....I truly am sorry" ("Relief from the Real Problem," p. 110).

Being Used

Some of the saddest stories we received were from gamete donors whose relationships were ruined after agreeing to help infertile or gay friends conceive. The donors felt used, lied to, and abandoned. Some endured dramatic health consequences:

> I gave my eggs away. A few years ago, I gave my eggs to good friends with the naïve belief that this was a way for me to be part of a family—and I would even have a genetic connection!...After the egg retrieval, [the attitudes of the doctors and nurses] towards me changed. They did not want anything more to do with me, especially when I had complications. I became an annoyance....I will be on drugs to control my hormones for the rest of my life. I still go months without a period. My mental health continues to be an issue and is very much linked to my hormones. My ovaries have remained enlarged, and I will most likely

> never be able to conceive naturally, if at all....My friends have decided that I am no longer part of their family ("Don't," p. 95).

Some contributors write about a process of seduction where social parents treat donors with great friendliness and warmth in an effort to make the donor agree to let the couple use his or her gametes. However, this period of warmth ends quickly after the couple gets what they want—the baby. Social parents often use diminutive language such as: "fairy god-donor" or "gestational carrier" and "funcle." "Donor" is still the most common euphemism used, though many believe that "biological/genetic mother/father" would be the more accurate and less confusing term for the child. Naturally, anonymity and diminutive language offer security to the social parent whose genetic role is being replaced by the donor. It makes for clearer barriers from the parents' point of view.

Notes

1. Melanie Thernstrom, "Meet the Twiblings," *New York Times,* December 29, 2010, http://www.nytimes.com/2011/01/02/magazine/02babymaking-t.html?pagewanted=all&_r=0.

9. Themes from the Fertility Industry

I am sorry to say that we have only received two stories from members of the fertility industry so far. I deeply appreciate the two contributors, but of course I cannot glean any insights from their entries alone. I warmly and authentically invite members of the fertility industry to participate in this discussion and explain their values and concerns.

With the lack of regulation and consistency in laws surrounding artificial reproductive technologies (ART) in the United States, parents, donors, and offspring currently depend on the industry to understand the seriousness of their business and regulate themselves.

But with clients in wealthy countries willing to spend hundreds of thousands of dollars for a single child, expecting the industry to autonomously act with great integrity may be gravely naïve of us. Still, let us ask the industry itself what areas we can improve so that the system functions better for everyone.

10. Themes from Parents

One would assume a reciprocal sensitivity between the emotional needs of all parents and their children, but I think that use of donor gametes puts an uncomfortable tension between parents and the offspring they bring forth. I think many parents begin holding their breath at the first announcement of a positive pregnancy test.

Feelings change as time passes. The perspective of the parent of a young child carries a notably different tone than the perspective of the parent of an older or adult offspring. One thing for sure, the decision to have a child through 3PR is practically never a first choice, and the road is long and complicated. Here are some common themes I gleaned from our authors:

No One Chooses His Family

A common argument in support of donor conception is the fact that no child ever gets to choose which family she's born into. Adults' wants will always trump children's, and children, especially the unborn, will always be voiceless and helpless. This is just the way things are: "All parents considering starting a family are thinking in large part about themselves, and no child, however conceived, gets any say in the matter.... Unborn children simply do not get a chance to approve applications from their prospective parents" ("A Real Dad Speaks" p. 150).

This is true. However, children *do* grow up and develop the capacity to voice their opinions. The intuitive parent will glean wisdom from these adult donor-conceived persons' personal accounts, take them to heart, and implement their advice. In so doing, the parent may avoid resentment from their adult child, and this is best for everyone.

Parents Desire Gratitude

Having a child through third-party-reproduction is often even more labor and resource intensive than having a child traditionally. Parents go through tests, screenings, and evaluations. They are prodded, pricked, and drugged. They must deal with guilt, stigma, and grief of the loss of a biological connection. Then when the child is born, they must meet the needs that all infants, children, and teenagers demand—from bottles and diapers to bass lessons and driver's licenses. In short, it is *not easy* to be a parent of a donor-conceived child. Naturally, it can be antagonizing for a parent when their donor-conceived offspring complain about their conception. With all the collective sacrifices a parent makes for a child, it's easy to view many of the stories submitted by donor-conceived contributors as ungrateful. One parent said, "[I]t truly bothers me to see some of the posts on here about how parents make unethical choices to conceive a child via donor....I hope that we will not raise a person who will make snap statements about people's life choices without knowing the entire story" ("A Positive Story—for a Change," p. 145). Another shared her irritation at being questioned:

> The question has arisen lately about why I would even consider bringing a child into the world where I knew she would probably never meet her biological father. And my question back is, "Would you rather not exist at all than to exist as you are?" Yes, I would do it all over again and bring you into the world, given the same set of circumstances that existed then ("Chaos of Emotions," p. 153).

Baby Cravings Are Powerful

Many of the contributing mothers write about how they "wanted that baby so badly." They say they tried to have the perfect family, but because of their bad dating record, their partner's infertility, their sexuality, or their age, they just couldn't make it happen. It is clear that baby cravings are a huge force in people's lives that inspires them to spend thousands, even hundreds of thousands, in the pursuit of parenthood. Some even divorce or threaten to divorce their infertile or unwilling partners if one does not support the efforts of the other to achieve pregnancy. I've experienced the same deep desire for a child, but I have been very blessed, both in love and in motherhood.

As a researcher, I suspect there is a major conflict between the ideal-life timeline women have been presented in the last few decades and the natural cycle of our bodies. We have been asked to suppress the primal desire to carry and raise children in order to close the achievement gap between men and women professionally. Many women are realizing too late in life that they're no longer able to get pregnant. And sadly, many of us have not been given the tools to establish healthy relationships with the opposite sex to succeed in marriage and coupling. Finding the "right" man seems outrageously difficult. *Keeping* him is a flat miracle.

But it is clear that cravings for and obsession with babies become just as much of an antagonist in women's lives as cravings for and obsession with sex are for many men. We expect men to get a healthy grip on their sex drive so that they avoid actions that undermine others' freedom, health, and dignity. We may now need to develop cultural scripts that encourage women to control *their* own brand of sexual/reproductive cravings.

Just as sex is a beautiful and healthy thing that can be corrupted and twisted into something evil, so perhaps it is with baby cravings.

Obfuscated Language

Parents use new words, often suggested to them by fertility-industry professionals, to help them make sense of their brave new relationships. The terms "biological mother and father" are intimidating to embrace, because they may undermine the kind of barriers intending parents wish to put forth:

> We also had some counseling before we conceived, through our IVF clinic, and the counselor took my breath away when he said, "Do not think of the donor as a father. Your child will not have a father; he will have two mums." Until that point, I had been scared about applying the term "father" to this man, because we knew he would play no part in raising our boy day to day.…But hearing that perspective really helped ("The Best Policy," p. 142).

Putting clients at ease is a natural and necessary facet of capitalism; the salesman who can achieve this will do his business a great service. Language is a great tool in directing emotion, and the fertility industry is clearly succeeding in choosing language that encourages the use of their services and products. Less cynically, it really does bring parents more comfort when they give themselves permission to graft new, more authoritative titles that replace the bio-dad/bio-mom norm. A social parent is much more likely to take her responsibilities seriously (which benefits the child) when she views herself as a true mom, rather than the biological parent's partner or secondary caregiver.

The Comfort of Epigenetic Imprinting

Epigenetic imprinting is when gene expression is changed and influenced either by drugs or environment. In this context, it is the quality of the womb and the decisions of the birthmother that affect to some degree which of the gestating child's genes are triggered.

Social mothers via egg donation often talk about the power of pregnancy and gestation to affect the development of the child biologically. Their accounts underscore the importance of their contribution by referencing the powerful bond between a woman and the child she births. They speak of breastfeeding, the choices they made while pregnant, and the "switches" they were able to turn off and on epigenetically while their child was developing inside their wombs. This potential for imprinting seems to be a comforting theme for social mothers via egg donation as they assert their authority and find security in their title of Mother:

> I also know that this child has literally been built from my body....His every cell has grown from my blood. Every choice that I made while pregnant—what I ate, when I rested, how I played—had an effect on his development, helped to activate or deactivate certain genetic qualities. This dynamic is called "epigenetics," and it means that my choices while pregnant influenced how my son's genes will be expressed ("Why We Chose Donor Egg," p. 149).

Fear of Invasive Donor

The popular use of anonymity in gamete donation is most likely attributable to the pervasive fear many parents have that the donor will somehow make his or her way into their lives, causing upset, confusions regarding loyalty, and jealousy: "Although grateful for their kindness and good intentions, I simply did not want a donor who might drop in to visit me and look at my child as if he/she were their own" ("Access Denied," p. 129).

Mixing one's genes with another person's is special indeed. There is a type of harmonious resolution when we see a child with its two biological parents, in love. Hollywood has delivered several movies regarding sperm donation, like Jennifer Aniston's *The Switch* (2010), and the happy endings always consist of the biological parents overcoming their conflicts so that they can be together as a complete family with the child. My mother was offered "help" from male friends as she and her (then) husband struggled with infertility, but she ended up not using a known donor because she thought if there were to be any marital problems, she might naturally be inclined to wander into the arms of her child's biological father. Anonymity seemed to solve that moral dilemma.

Consummating a Second or Late Marriage

The divorce revolution ignites demand for 3PR as couples wish to fully consummate their new love through a new child. Eggs from younger women are used as people remarry later in life, often after a wife's fertility has steeply declined:

> I was in a new marriage, with a wonderful husband and two gorgeous children from my previous marriage, and I was on the threshold of acting upon the yearnings to have a child with my new husband, especially as he was such an amazing stepfather. I felt he deserved the right to have a child of his own ("Access Denied," p. 129).

Men, having no age limit on their reproductive capacities, have never historically rushed into marriage. In recent history, we have stigmatized older men who wait until late in life to then marry younger women. Picture the weathered old gentleman who carries a twenty-three-year-old on his arm, having kids with a woman young enough to be his daughter herself. So today, men are still waiting until their later years to settle down, but they've listened to the social pressures to date women their own age. This has increased the demand for donor eggs and surrogate wombs: "We were older and knew my ovaries were unlikely to do the job, but he came to me without any children of his own, and he wanted kids, and I love kids, and so we decided to try to have kids" ("We Have Children…," p. 139).

Gender Wars: Better to Have No Dad than a Bad Dad

I wasn't surprised to read stories from women who chose to go ahead with donor sperm because they saw no prospects for finding a good and decent father for their child, having been raised by poor fathers themselves and suffering from disappointing relationships with men as adults. One contributor writes, "I never knew my real father. He left when I was three and my sister was one.…At least my child won't feel the hurt I felt knowing that their dad met them and still chose to leave" ("Optimism and Confusion," p. 136). Another shared, "My father was always quite critical and even stated on occasion that he'd never wished to have children. This devastated me. So, you could say, I did not see the fact my daughter would not have a father as an entirely bad thing" ("Thankful for My Donor," p. 141). Another was even more blunt about her regrets:

> I was the child of a womanizer. My mother fell in love with a man who would never grow up, and I paid for it.…I grew up and fell in love, but always with abusive men.…I think that I could have been much happier if I had not had a dad like mine. Sincerely, I would prefer to not have had any ("A Lifetime Yearning," p. 137).

The gender wars show up often in stories of donor conception. When men and women don't trust each other, they don't marry, and they don't try to cooperate for shared and equal access to their children. In the book I coauthored with my husband, *Happy Couple Creed: How to Make Your Love Last a Lifetime,* we have a chapter titled "Love Needs Trust. Trust Needs Virtue." This concept has implications for the entire economy. When people don't trust each other, they don't cooperate or share. In order for trust to grow, people must demonstrate trustworthy behaviors, or virtues. If a culture does not promote the seven virtues of chastity, temperance, charity, diligence, patience, kindness, and humility, then it will self-destruct as disappointing behavior leads to mistrust and noncooperation.

The Parental Love Is Real

Undoubtedly, there are strong feelings of affection and responsibility for the child from the parent(s):

> I love my son. He already knows about his origin, and although at four he is not that interested yet, he will always have known his story and he will never feel deceived. Also, he and I are very close and loving, so he will never feel he was missing a father figure.…I do not apologize for choosing to bring this wonderful person into the world, and I would certainly do it again ("A Real Dad Speaks," p. 150).

"The parental love was and is strong and did not nor will not change through the years," one mother shared. "She is my child, and I am very proud and happy to be her mom" ("Chaos of Emotions," p. 153).

I wonder if love (such a fluffy word) really is enough. In many cases I'm sure it will be. In others, I have my doubts. Life has a way of balancing things out. I remember my mother once telling me she didn't want to buy a newly designed car because she didn't want to suffer from "all the kinks not being worked out." She opted for the tried-and-true Honda Civic instead. That's how I see marriage and reproduction. Embracing the "tried and true" social-engineering advice passed down from the centuries (like fidelity and life-long commitment) seems to gift the user with stable relationships and usually, children. With 3PR, we're trying to build a revolutionary new engine, but we're still working out a lot of "kinks." But just as love built the Taj Mahal, it will be the tinkerers and troubleshooters who are truly filled with a sense of love and dedication that will promote the best practices and successfully serve the needs of children and parents in tandem.

11. Themes from Partners, Friends, and Others

Third-party reproduction affects others in addition to the obvious players of parents and the donor-conceived. It affects the future spouses and future children of donors. It affects boyfriends/girlfriends/spouses of donor-conceived people. It affects our communities. It affects our families.

Donor-conception affects how we frame our lives—our motivations and logic—and it influences the decisions others make and how they frame their own lives. As a culture, we decide what is justice and what is justifiable. These are the stories of secondary players. I want to thank them equally for their participation.

Go After the Law

Several contributors have encouraged donor-conceived people to pursue a change in the law. They encourage the active channeling of energy and resources into the support of our common interests. For instance, "I hope that all of you children grow up and write to your government officials to change the laws regarding anonymous egg/sperm donation in your state. You children do have the power to change the laws so that anonymous egg and sperm donors can no longer be anonymous" ("My Story for You…," p. 158).

I agree that there will never be change unless change is demanded. Complacency doesn't catch the ear of power-yielding lawmakers. How will anyone know there's a problem unless they are notified?

> There are several countries that allow people to access their donor's identities once they are eighteen years old. The U.S. is not one of those countries, but that's o.k., because what the hell kind of country responds to human-rights violations by asking its disenfranchised people, "What can we do to make being discriminated against a more pleasant experience for you?" Donor-conceived individuals in the U.S. will, I hope, begin to compare themselves to those they want to be treated like so they

> can identify what laws need to change to make that happen ("Did You Know You Were Supposed to Be Equal?" p. 160).

Empathy from The Fatherless

People who grew up without their fathers seem very able to empathize with the donor-conceived experience. They see their situations as related to ours and offer their stories as tangential supplements:

> I asked who my father was. I yearned for my father. My gay mother told me I did not have a father. I persisted and insisted that she tell me the truth.... [My father] died of a hereditary illness four months after I found him. I cannot tell you the closure and peace that came from just meeting him ("My Story for You...," p. 158).

A "Parent's Perspective"

One woman wants to speak up on behalf of all 3PR parents:

> As a potential parent of a donor-conceived child, I beg you to try to see it from a parent's point of view. You are absolutely entitled to be annoyed at your parents for hiding it, for not knowing your biological parent, but please know that it can't have been an easy choice to make for them ("It's Scary to Need a Donor and Read These Stories," p. 165).

As with most topics, as children, it is nearly impossible to *avoid* seeing donor conception from a parent's perspective. It's the first and solitary perspective we have for a while. After all, we spend our entire childhood in the same home as our parent(s), listening to and learning from them. Basically all donor-conceived people are pro-donor-conception as young children. If we ever become critical, it is in adulthood, often late adulthood.

It can come as a great surprise to parents when their adult child first vocalizes anti-donor-conception opinions. This shock can lead to confusion and hurt, especially when parents feel they have done everything right for eighteen-plus years.

Moral Dilemmas

This world of third-party reproduction is ripe with intellectual challenge, as people attempt to balance their desires against their conscience. It's not just parents who struggle to "do the right thing," but their future spouses, friends, and extended family. Indeed the world of infertility is an effervescent chamber of moral dilemmas—where combustible elements with different energies and goals meet together, sometimes in extreme heat:

> I don't know how to tell my children that their half brother has in fact got another half sister who is nothing to do with them....I don't know

> why I find it so hard.…I guess I have to encourage my partner's relationship with the little girl he fathered in this way. It's not her fault. I know all this. But it is hard ("What It Feels to Be the Partner of a Sperm Donor," p. 155).

There is, of course, much more to be said about third-party reproduction and the real-life challenges it produces. The Japanese have an expression: "As the radius of knowledge expands, the circumference of ignorance increases." Similarly, we have only begun to discover what we didn't know we had lost.

Here begin the stories from our contributors. I kept the original titles as contributors wrote them, and the stories have only been lightly edited for clarity and convention. All ellipses (…) were included by contributors; they do not indicate omissions.

12. Personal Stories from Donor-conceived Persons

"I found out a little too late."

I may be a little bit unusual as far as the children of sperm donors go. I am a fifty-year-old woman, and I found out exactly one week ago today. It came from out of the blue, because I have done genealogy for the last twenty years and very recently decided to take a DNA test. I think in a way that I am still a little bit numb, but when the feelings come, they are strong and painful. It's always on my mind. Always. The worst part that I see so far is that no one in my circle of family and friends even has the slightest clue as to what I am feeling. I should put an exception on that though, because despite my parents' wishes, I told my grown daughter. She was nearly as crushed as I was. We identified so strongly with my dad's Scottish heritage. It was part of who we were. I have not told my son yet, as he is away from home in the military, and I know this will crush him. I want to tell him in person.

Until then, I will wait for my DNA results to come back and hope that I can find a few clues as to who this man was. My mom only had very vague memories of what the Dr. told her when he did the procedure. I was able to find out the name of the Dr. and find some of his background information online, but he died about fifteen years ago. I know my parents only wanted children, and that maybe, at best, the donor only wanted to help someone. The unfortunate part is that now, all the pain and misery is heaped on me. Knowing what I know now, I strongly believe that sperm donation *has* to be an open procedure; that the child should have full access to the donor's information. It's the only moral thing to do.

"My Forgotten Half"

I was born in late 1985 to a single mother, a nurse who had never married but had wanted a child for as long as she could remember. She saved up money and went to a fertility doctor, somewhere in the Bay Area. Twenty-five years later, my mother suddenly died from a stroke in her sleep.

As a child, I always knew how I was conceived, or at least knew the basics. My mom never hid that from me. I was raised Catholic (though not too much so) and wasn't baptized until I was five. I guess my mom was afraid I would be rejected by the Church, but the Catholics in the Bay Area tended to be a bit more progressive. For me, not having a father was a normal part of life; I knew nothing else. It was just me and Mom, no siblings, at least that I knew. When I was twelve, my mom decided (for reasons I'll never quite understand) to pack us up and move to southern Oregon. Seventh and eighth grade there were nothing like previous grades in California. Where in California I had been open about my origins and never thought there was anything wrong with where I came from, the two years in Catholic school in Oregon were hell. Instead of having a name, I was simply "test tube" and never could get a moment's break from it.

When my two years there were finally up, I made a choice. From then on, I was simply going to tell everyone that my father died when I was a baby and I never knew him. I told my mom about this, and she understood but was saddened by it. It wasn't easy to hide where I came from, but over all, it made high school easier.... Well, less horrible than it would have been otherwise.

One day when I was seventeen, I was talking with my mom about what it would have been like if I had grown up with siblings. With that conversation came something I had been kept from for over ten years. It turned out my mom had managed to get pregnant again after I was born, when I was five. She was going to have twins, but before they were fully developed, the girl died. Shortly after, my mom gave birth to the boy—who didn't last long. It was strange to know that for a brief moment I had a little brother. It was sad, but probably for the best. Growing up with only one parent that was mentally ill hadn't been the best, the easiest, and chances are I would have ended up a parent to any siblings.

Once I was out of high school and able to separate myself more from small-minded idiots (there's a lot of them in southern Oregon), I was able, once again, to be honest about where I came from. The few high-school friends I had left were amazed but supportive. Many new people I met were fascinated. As time went on, though, I thought more and more about it. Not so much about my father, but the half siblings I had out there that, chances were, I would never meet.

Last year, over spring break, my life took a huge turn. I was on spring break from college. My mom had been ill for some time, and her doctor had no idea what was wrong. She had dizziness, confusion, and times of total incoherence. I had taken over doing most of the things she used to do herself, or at least anything that involved dealing with other people or driving. I had been doing this for over six months at this point and was ok with it. She had raised me alone, and I felt this was my chance to repay some of that. Over spring break, she had been particularly bad. I was due to go back to my college to help a classmate film something. That day, though, she had been feeling especially bad, so I decided to stay. She had been napping all day on the couch, and that night, when I came out of the shower, I found her not breathing. Despite my CPR attempts and that of the paramedics, she was gone, taken out by what was most likely a stroke.

Now, a year later, I find myself wanting to find either my father and/or siblings. It would be nice to not feel alone. I have family, and they've been great, but they're all 500 miles away. The sad truth, though, is I'll probably never locate any siblings. Doesn't mean I won't try, though.

"OK"

First they said we would be o.k. because at least we'd know who are mothers are.

Then they said we'd be o.k. so long as our parents tell us we were donor conceived.

Now they say we'll be o.k. so long as our parents tell us we are donor conceived and we can access the identities of our biological donor parents.

When are they going to work it out that we'll only be o.k. if they admit that donor conception is not o.k.?

We'll be o.k. when society recognises the hypocrisy of recognising the importance of biological familial relationships and then saying they aren't important if you're donor conceived.

We'll be o.k. when we are permitted to grieve that loss. To say it out loud. And have open ears receive the words without retribution.

o.k.?

"Wondering"

Every day I wonder who you are. If I will ever get to know who you are. Even a picture. If you hate mayonnaise as much as I do. I wonder if you ever think of me, or any of us. Every day hundreds of questions circle through my mind, and none of them are ever answered. I want to know who you are, and I think I have a right to at least know that.

"Meeting My Donor-conceived Half Brother"

A few years ago, I was put in touch with my half-brother, conceived using the same donor that I was, through the UK voluntary contact register for people conceived before 1991 (UK Donorlink).

Initially it was so exciting. Close in age, we emailed each other most days, keen to catch up on all the years we'd missed together, soon graduating to long phone calls and one meeting in person. I felt we were essentially alike, that we had a bond and he would always be in my life. It even prompted me to tell my (social) dad that I knew I was donor-conceived, even though I'd promised my mother I would never tell him, because I believed I would not be able to hide the relationship from him forever.

But soon cracks began to appear. He would ring late at night and want to talk for three or four hours. His deep distrust of people started to show. He was jealous of me spending time with other donor-conceived people, convinced I would like them more than him and ultimately reject him. No amount of reassuring him that I would try to accept him unconditionally helped. He confessed to violent fantasies about hurting his social dad. These calls were draining and depressing. On my counsellor's advice, I limited them to thirty minutes at a time. He didn't like that and became quite abusive, accusing me of being emotionally repressed, for example. Gradually I stopped answering the phone.

In the last email he sent me, he said, and I quote, "People make me sick. They are dull and aggressive, even toward each other." I realised it is impossible to have a relationship with somebody who is starting from the position that you are probably a psychopath.

It may come as a surprise that I do not regret making contact with him and I still regard him as my kin. But whilst I'm still looking for other half-siblings, I would be more guarded now if I were put in contact with one. Me and my half-brother ARE alike, but what manifests in me as social anxiety manifests in him as something darker, and I would have to wonder if anybody new I met was the same. I've learned the hard way that kinship does not necessarily equal friendship.

"Uncertainty is Killing Me"

Who are you?
Will I pass you in the street?
Will you hold the door for me and smile as I walk into a gas station during my travels?
Will you look at me and wonder if I belong to you?
You have the pleasure of knowing I could be here.
For nineteen years, I was denied of knowing you even existed.

What was *my* grandmother, *your* mother like?
What about *my* grandfather, *your* father?
Why do you get to selfishly keep them all to yourself?

Who are you to deny me half of my family tree—
Branches rich and strong with stories I may never be told?
Who are you to give away my heritage, knowing it will be replaced with something false?

Do I have brothers and sisters with my dark hair, my deep brown eyes?
Will I be attracted to a familiar stranger in my classes?
Will I fall in love with him and kiss him passionately in an act of accidental incest?

Have you told your wife?
Have you told your partner?
What about your children?
Have you told your brothers and sisters about their mysterious niece?

Are you dead?
Will you ever read this?
Have you dismissed it as something in your past that you did to make ends meet?

Did they pay you to give me away?
What did you spend the money on?
Did you buy a sparkly necklace for your ex-girlfriend?
Did you buy books?
(The bank you went to would have paid you half of my College Algebra book for the donation that included me).
Did you buy a candy bar at a gas station?
Was I worth it?

Do you miss me?
Do you ever think of me?
Do I even cross your mind?
Does the uncertainty drive you crazy?
Was it worth it?

Do you wonder when my birthday is?
What color gown I wore to my graduation?
Would you be proud to know I was the valedictorian of my senior class?
Would you support that I am Christian?
Would you even want me in your life?
I want you in mine.
I will accept anything about you, if I could just get the privilege of knowing who you are, of knowing who my family is.

"A Response to 'Donor' "

[This is a response to "Donor," found in this volume on page 101.]

Dear "Donor,"

I am so sorry for the position you now find yourself in. But you are all of your offsprings' father, no matter how they were conceived. There really is no such thing as a "donor" from your children's perspective. Your wife, parents, brothers, sisters, nieces, nephews, and especially your children (all of them, no matter how they were conceived) all deserve to know about each other. I'm glad you now see the error of your ways, but denial will not redeem you and in fact might make things much worse in the future. It would be better to be honest with your family now, before one of your children, who you did not agree to raise, finds you through DNA testing. It's too late.

"A Letter to My Father"

Who are you?

I ask myself this question every time I catch a glance of my reflection in the morning, and not just because my hair is an absolute state.

I fear, fear for what the future holds and wonder if I will receive any slight information about you.

I grieve for you and for the part of myself that I will never truly know. I feel as though half of me is missing alongside my family, because that's what you are. Family.

I feel as though I have been cheated out of knowing you, communicating with you, as I never got a say in the first place. I still don't now.

You are Anonymous.

And there is nothing I can do to change that.

Who are you?

Do you honestly care about the poor couples out there unable to conceive?

Or did you just need the money? Do you want to know who I am? Do you think about me at all? Are you gay?

All these questions are left unanswered. I just want answers, such a simple ask, but one that will never be fulfilled.

I'm transgender. ftm, that's female to male.

I have not told my mum or my dad for that matter, because I know for a fact they would be ashamed of me.

Would you?

I talked to a psychiatrist, told her my worries. Apparently not knowing who your biological parent is can mess with people's brains. My parents will blame themselves.

Who are you?

You must have brown hair, blue eyes, and be around 5'8". Just like my dad, that's how it works, right?

That's all I know.

How many other half siblings do I have?

I know my other sisters, twins (but due to two embryos being placed in my mum's womb), but what about half. There has to be some, is somebody at my school related to me? I will never know.

Do you pass me in the street? Do you do a double take between the strong resemblance I have to your sister? Or do you not even notice.

Are you smart? Do you play any instruments? Do you dance or act?

Do you get worried, never for yourself but for the thousands of others around you? Do you have a total disregard for your own safety?

I do.

Nobody in my family does.

Do I get these traits from you?

I hope so.

My sisters don't seem to care at all. They never talk about it and act as though everything is fine. It's not.

I notice things more. Always have I suppose. Apparently nothing gets past me, don't know whether that's a blessing or a curse.

I began to think about you when I was eleven. My best friend's mum died of cancer; she had been battling for two years and lost the fight just after my friend began secondary school. I was left wondering if you had died, if you were dying and if I would ever get the chance to see you.

When I was fourteen my granddad died. He was my dad's dad, and my world came tumbling down. I loved him to pieces and really I should have seen it coming. He smoked like a chimney; let me have a sneaky smoke of his cigarette when I was seven in the back garden. It then came as no surprise when his prostate cancer spread to his lungs and turned fatal.

I was young and took everything for granted. I knew deep down he was going to die, but it still came as a massive shock.

I was guilt ridden. When I found out I was donor conceived, I began to see Granddad in a different light. I used to pick up on tiny things that now I realise meant nothing and turn them into something else. When he passed, I realised how stupid I had been and was angry with myself. For not treasuring those last weeks I had with him.

I began to self-harm. It was the only thing I could control. I felt like it was a release, from the messed-up life I led, to a new exciting experience that I was controlling.

Just before my fifteenth birthday, my other best friend's dad died. He went away on a holiday to Rome with his best mate Greg and never came back. Heart attack at forty-one.

I was lucky to have a dad and a mum, forget biological or not. I still had two parents. She reminded me of that and how lucky I was.

Who are you?

I need to know. After all these years.

I wish I knew who you are.

When I was little, I used to pretend you were someone off of the TV. A celebrity or even a rock star. I would make stories and play them over and over in my head, sometimes writing them down and I would always dream about them.

I say little, but this "stage" lasted well into my teens.

If only I could speak to you, write a letter. That's why I have written this, hoping that by some miracle you may come across it.

You wouldn't know, though, that it's you I'm talking about; you would pass it off on some other guy. Perhaps I'm making you curious, or maybe you know for sure that this letter is not for you.

Or do you?

I know for a fact that if I ever get the chance to meet you, what my first question will be.

Who are you?

"The Cup"

A dialogue between a sperm donor and his child who don't know who each other is:

I saw her pain
On the paper and TV screen
Anguished tears
Of failed familial construction
A child lost
In the depths of minds conception

She felt a need
The need to hold me in her arms
She felt the yearning
Of every woman so it seems
Who wants to be
Maternally in their dreams

Here's my cup
See it is half full
I gave you away
Deception will rule

Here's my cup
See it is half empty
You sold me away
Completely abandoned me

Here's the cup
We swim in the sea
Of men and boys
No roots for the tree

Screwed on the lid
Of the lives I'll never know
Golden handshake
Of the human industry
How was I to know
Of the pain you won't be free

Now I'm alive
Required grateful for existence
Wanted so deeply
Eternally burdened and indebted
Identity crushed
How it must be, is how they said it.

Here's my cup
Now it is half empty
I threw you away
Will you forgive me

Here's my cup
It is still half empty
Of kinship lost
Could you not see

Here's the cup
We swim in the sea
Of men and boys
Traumatised we'll be

My child once lost
Will never be regained
Our blood once severed
Created scars too deep
For this I am sorry
Your forgiveness, you can keep

Dehumanised I am
A commodity of modern marvel
An object of desire
For adult tear jerked masses
They took you away
My father in the street passes

"I Have Never Met My Donor 'Father,' and I Have No Desire to Do So."

(Submitted to the Law Reform Committee, October 2011, Australia)

I was conceived through sperm donation, and my mother was adopted, so the nature and meaning of genetic relationships is something that has long been thought about and felt deeply in my family.

I have never met my donor "father," and I have no desire to do so. I do not see this lack of contact with my biological father as something missing in my life, and I have no hurt at the fact of my creation.

What does cause me hurt, however, is the idea, constantly repeated by small numbers of donor-conceived children, and in popular media representations of the issue, that there must be something wrong with your life if you do not know your biological father.

Over and over again we hear the phrase that there is "something missing" in the lives of donor-conceived children, some meaning in their life that has been taken from them.

But we all know what it is to question our place in the world. Most people eventually develop a sense of place and personhood, or at least learn to live with feelings of uncertainty.

Yet for people who are adopted or born through donor egg or sperm, society continually reinforces that feeling. We are told repeatedly that yes, there is something missing.

This is a pervasive and cruel cultural myth, this idea that you cannot know who you are unless you know your biological parent.

My whole life I have been subjected to the shocked responses of people who on learning that I don't know my donor could not understand how I could be comfortable with this and believed I must be harbouring hurt about it.

They found it difficult to imagine life without their own father, and they did not differentiate between their biological connection and their emotional, familial connection born of shared experiences and values.

But if I ever meet my biological father, I will not find a "father" in him. He had no input into the adult I am today, besides his kind gift of the genetic building blocks I needed for my life to begin.

The reality is you—the things that make you really you—are not your genes. There is no gene for your love of camping, or the comfort you get from the smell of your mother's perfume.

Knowing your biological parent will never explain why you love the one you do, why you hate early mornings or feel uplifted by classical music. Genes interact with your environment to create you, but they do not give meaning to your life.

Even people who have a very similar genetic makeup can be very different, and science is still far from tracing genetic roots for most illnesses, including those more linked to our personalities, such as some mental illnesses.

And the circumstances of our upbringing further pollute our genetic stories. Which genes are expressed in us turns out to be a complex, messy mix of environment, culture, and perhaps even chance. The idea that on learning about a donor you would somehow be given some important insight into your "self" is, frankly, ridiculous to me.

Of course, this is a trick we play on ourselves every day when we light-heartedly explain our personality traits or those of our siblings as being "in-herited" from others in our family. Normally it is completely harmless, but in the case of donor-conceived people and adoptees, it creates a completely unnecessary and often damaging idea about what being donor conceived means for their life.

When my mother first met some members of her biological family who lived in the country, she came home and said to me, "Now I know why I love the bush." Her strong belief in biology meant she couldn't help overlooking both her history with her adopted family that could have led her to love the bush, and differences with her biological family which did not fit this narrative (such as that she could never bear to kill an animal or really do anything other than coddle it, but her biological relatives, who were farmers, by nature of their jobs had a different attitude).

I note with sadness a submission received by your committee from a young woman who had grown up in a Maltese family, but it was not until she discovered her donor was Maltese that she allowed herself to feel a connection to that country.

What a tragic and fruitless search is being pushed onto donor-conceived children by the idea that the only place they can find such personal and cultural meaning is through a biological connection rather than cultural and personal connections.

As a society, I believe, we should be working towards developing a more mature and meaningful debate about what role biology plays in who we are. Rather than pandering to quasi-religious, biologically determinist beliefs, we should be actively challenging such ideas.

As the biologist Stephen Jay Gould said, each individual has the potential to be and act in many different ways. We have "a brain capable of the full range of human behaviors and predisposed towards none." This conception of humanity is clearly more nuanced and realistic than a simplistic biologically determinist model where we all have specific genes for our specific behavioural traits.

The idea of genetic determinism traces back to thinkers of the nineteenth and early twentieth century and has long been replaced with far more nuanced ideas in the scientific and philosophical literature. It's high time the rest of us caught up.

Arguing that not allowing donor-conceived children access to their donor violates the United Nations Convention on the Rights of the Child is based on faulty assumptions about the role of these biological fathers in their lives. Three of the Articles of the Convention mentioned in the draft report are:

Article 3 (Best interests of the child): The best interests of children must be the primary concern in making decisions that may affect them.

Article 7 (Registration, name, nationality, care): All children have the right to a legally registered name, officially recognised by the government. Children have the right to a nationality (to belong to a country). Children also have the right to know and, as far as possible, to be cared for by their parents.

Article 8 (Preservation of identity): Children have the right to an identity and an official record of who they are. Governments should respect children's right to a name, a nationality and family ties.

These Articles should not be used to justify such a change. Propagating these cultural myths is not in the best interests of the affected children. These children ALL have existing rights to names, nationalities, parents, histories, and identities. They have families who love them, raised them, and in doing so shaped who they are today.

To think that the very real rights of men who donated anonymously and do not want to be contacted by children could be violated on the basis of cultural mythology that I believe we should be fighting, rather than endorsing, saddens me greatly.

However, I should note that not all donor-conceived children feel the same way I do, and I do not intend to undermine the absolutely real and very strong emotions they feel about not knowing their donor fathers. We as a society have created this mess, and we should fix it up. I would support an option of paying for genetic counselling for such children, as well as a mechanism for them to make contact with anonymous donors IF the donors also wanted that contact, as this does not place the desires of one of the two people above the other.

In conclusion, I do believe that many donor-conceived people suffer a real and tragic hurt from the knowledge that they do not know their biological father. I think the government should undertake to attempt to contact donors if approached by donor-conceived people, as it will likely cause little individual harm if both donor and child want to meet.

But the hurt felt by donor-conceived children is largely based on misinformation and misunderstanding, and to further compound it by changing the laws and allowing men who identified anonymously to have their identities revealed is the wrong approach. Not least because, if those men have already stated that they do not want to meet, revealing their identities will likely not provide the children the connection they desperately want. These children should be supported with counselling services.

I also hope in writing its final report, the Committee will acknowledge that the "genetically determinist" ideas about sperm donation are harmful in and of themselves, and will challenge these ideas rather than accepting them.

"For Nineteen Years Now..."

Let's see....When I was six, my parents got divorced. It was rough, but my dad was going through a midlife crisis, I guess. He was in his midforties and decided to recapture his youth with a young woman. He dated lots; my mom struggled to raise me by herself. I always found myself desperately wanting his approval and love. When I was eleven, my mom sat me down and had "the talk" with me, when she told me that my dad wasn't really my dad, that I had been conceived in a clinic.

I don't know if it was the fact that I always wanted him to be a dad to me, or that I was only eleven years old, which is a super-big transitional time in a young girl's life, but the news was devastating. I felt like a freak of nature, a science project gone awry. I began to doubt every single thing about myself, every emotion, every thought, and chalked them all up to an error in the freezing/cryogenic process.

Many years have passed since. I am now a thirty-one-year-old woman, with three kids of my own. And I see pictures of myself and wonder, what does my biological dad look like? People say I look like my mom, and my dad (always trips me out, since he is not my bio father), but what kind of genetic/hereditary traits do I get from him? Who is he? In 1978, modern-day medicine was not at all what it is today. What did they screen for? Am I prone to cancer? Am I prone to heart disease? Hypertension? When I take my kids or myself to the doctor and they ask my family history, I always have a hard time saying, I don't know who my biological father is. All that I know is that he was supposedly a medical student at UC Berkeley. Is that even true? Is that why I love medicine, anatomy, physiology, etc....Is that why I'm freakishly intelligent?

I want to see a picture of him. I want to know if I have his laugh. I want to know those kinds of things. I don't want to be his best friend. I'm not looking for a new daddy. I just want to know where I come from. I think that is fair. Adopted children get to find out someday. Why don't donor-conceived children ever get to find out? It makes me sad to know that I will never know. For nineteen years, I have tried to come to terms with the fact that I will never know, but still, there is a hole that can never be filled.

"Growing Up Without a Father"

When I was young, I would create elaborate stories of who my father was. My friends would ask me about him, and I would say he died before I was born. There was a series of pictures in our hallway that had a sailboat tipping over and half a dozen men falling overboard. I chose the man with a full beard in a long, yellow rain jacket. I said that was my father and the picture was of how he died.

When I was in kindergarten, we had to create a poster of who we were. We were supposed to talk about our favorite color, sport, food, and our family. After I presented my project, a kid asked me why I didn't have a picture of my father on my poster. The teacher quickly changed the subject, and I went back to my desk thinking for the first time ever, why don't I have a father? That kid was the first of many questions to come. To this day, people still ask about my dad. I tell them and say that I don't have one. Most people get a confused look on their face and move on, but some go on to ask what had happened. I just say that he died, to save myself the breath and to satisfy their intrusive curiosity.

Until that day in kindergarten, I never found it odd that I had no father. My mother and I had a happy life together. She was forty-five when she had me. She had been married twice, but unfortunately, both of them passed away. She was getting old, and she wanted a baby. I don't blame her for finding a donor, but I wish she would have thought about what I would one day feel like. I hate not knowing who half of me is. I look and act nothing like my mother, and I often wonder if I am anything like my father. I daydream about meeting him. Would he have my sense of humor? Would he like my music? Do I have a half sibling? Even though it makes me sad to think about those things, I can't help it.

I do want to find my father one day. I like to think that it isn't impossible. I'd like to know where half of me came from. It's very hard to think that I have never been nor will ever be in the Father's Day section in the card aisle. It's even harder to think that I will never have a father to walk me down the aisle when I get married.

"I Love My Daddy"

Being donor conceived can be really tough if you never knew about it or never suspected not being your dad's "biological" daughter. Never did I have one thought in my mind that my dad was not technically my genetic dad. It wasn't until a few weeks ago at the age of twenty-two and a senior in college that my mother flat out told me when I least expected it.

I did not cry then, as I was in shock and did not know what to say. My mother thought I would be relieved to know that I wasn't biologically related to my dad, due to him already aging and showing signs of mental illness that I have also seen run on his side of the family. My mother and I are a lot closer than I am with my dad, but a lot of it has to do with the problems related to his aging. But he wasn't always this way; he used to be normal and healthy when he was younger, and he was and is a great dad.

My mother told me for all the wrong reasons, and she felt it would give me a confidence booster in college since I am a predental student. She thought if she told me that my biological father was a med student at Yale (where the procedure took place), therefore implying that I get my intelligence and interests in science and health field from my sperm donor and not my dad, that I was fully capable of excelling in college. Her telling me totally did not work in the way she intended. Instead I felt like I had been living a lie. All these years I was so sure of who I was, and the family that I had. I always felt so sure of being a kid with "normal" family. I even saw resemblances with my dad, like being tall, a big nose, and flat feet. I thought I had my grandmother's hands and my aunt's legs. Now I know it's all a lie.

What's even more ironic is that all these years, I have never directly called my dad "dad" but instead called him by his name. I wonder how much this has hurt him, reminding him I'm not biologically his, when I didn't even know so. For a few days, I saw my face in the mirror with curiosity, wondering what my donor looks like. Wondering if indeed some of my interests come from him. It's only curiosity to know more about myself, nothing more. I wonder what my other heritage is. I always thought of myself as fully Hispanic, but with my very fair skin complexion, I wonder if he was possibly a brunette Caucasian. Many questions that I wonder, that I never wondered about before.

As others have stated feelings of abandonment by their donor, I do not feel this way. I actually have no interest in meeting this man or forming a relationship with him. I just want an exchange of information for my own personal questions. I would like a picture so I can take away that curiosity of my physical features and compare my big nose to his and laugh. Other than that, I could care less, because I have my daddy, who has always been there for me. I have a dad, and he is irreplaceable, although he has his issues and may not have been an intelligent doctor. He has been the best dad I could ask for. Nor do I feel betrayed by my mother or my dad for not telling me all these years, as I feel this is something no one could ever say when is the perfect time to tell your children. I was very rebellious as a teen, and knowing this before, I could have said very mean things to my dad, like possibly saying, "You're not my dad anyway." So I'm glad I did not know when I was younger.

I realize they wanted a family and if it weren't done this way, my daddy wouldn't have a daughter or family. He would end up in a home aging with no family to watch over him. But because I exist and love him, I will never abandon him. As the flashbacks come to my mind, of how wonderful my dad has been, like taking me to get ice cream every Sunday, to my doctor's appts, always remembering the exact time I was born, picking me up from school, crying his eyes out and falling on his knees when I was hiding and they thought I was lost, bawling out when I had to leave for college, and even in my rebellious years of treating him like crap and not once did he respond, "You're not my daughter." That, my friends, is more than I can ask for in a dad.

He always treated me as if I were his own blood and called me his daughter and been proud of my accomplishments like any other dad would be. For me this has worked in a different way than for most. Finding this out now made me realize how much I love my daddy, and I don't care if I'm not genetically his. I do look like him in so many ways and even physically, due to all the love and nurturing he gave me my whole life. The saying is true, you don't know what you have until you lose it, and I hope everyone cherishes their dads and lets them know how much they mean to them.

I did cry for many days after I found out, but not because of myself, but instead I cried wondering the pain my dad went through to find out he couldn't have kids of his own. And I also wondered did he love me unconditionally as if I were his own? But once the memories kicked in, I realized he has done well and beyond proven his unconditional love for me. I love my mommy and daddy so much, and I am still the same person I was before I found out, and I could care less about the rest. I have my family and I love them. I am thankful for having parents who love me unconditionally and had all of the desires to have me. What was I bickering about? I DO have a normal and complete family!

"Endocrine Disruptors"

This is an environmental-policy issue. I was fucked over from having a relationship with my father basically because of pollution. My "social father" had a rare disorder where he had three chromosomes: XXY. This means he was basically male *and* female. He had a male body, but during puberty, the estrogen secreted by his female-informed endocrine system basically rendered him sterile. High amounts of estrogen (or estrogen impostors such as PCB and dioxin) can also cause a man to display increased promiscuous behavior—he was a frequent philanderer.

Endocrine disruptors morph sexual orientation, too. Here's an excerpt from *Our Stolen Future* by Theo Colborn, Dianne Dumanoski, and John Peterson Myers. "Small doses [of dioxin] did long-term damage to the reproductive system of males exposed in the womb and through their mother's milk.... The mother rats had swallowed only a single dose of dioxin on the fifteenth day of their pregnancy, a critical period in the process of sexual differentiation that causes males to become male and not female. As they matured, the male pups born to mothers given dioxin showed sperm reductions of as much as fifty-six percent when compared to those whose mothers had ingested none. Moreover, even at the lowest dose, the male pups showed a sperm count drop of as much as forty percent."

"Dioxin also appeared to affect the sexual behavior of male pups exposed early in life, suggesting that it had interfered with the sexual differentiation of the brain. At maturity, these males showed diminished male sexual behavior in mating encounters and increased propensity to exhibit feminized sexual behavior, such as arching of the back in a typically female lordosis response, when treated with hormones and then mounted by another male."

PCBs and dioxins are found in every mammalian species, on every continent, in every corner of the world. Here is a list of ways you can expose yourself: combustion, metal smelting and refining, manufacturing chlorinated chemicals, paper bleaching, eating animal fats and meats, full-fat dairy products, fatty fish, breathing incineration gases, handling chlorinated anything, making pressure-treated wood, operating with municipal solid waste.

Essentially, if you're a mammal and you have a metabolism, you have been exposed to endocrine-disrupting toxicity, and your immune system, physical reproductive capacities, and social/motivational reproductive capacities have been compromised.

Sperm donation is not a solution to low sperm count.

And using it to create *more* people to produce *more* pollution is also not a solution.

"Years of Feeling Conflicted"

When I was twelve, my father brought me into the living room. My mom was already sitting on the couch and crying her eyes out. My father then explained how he had had a vasectomy before he had met my mom and that he couldn't have conceived me (this was pre-1980). He explained, with my mother crying the whole time, how they went to a doctor who had, with the help of a donor, conceived me. At the end of the conversation, I asked my mom why she was crying so much, and she said, "I didn't know how you would react." It took so many years for it to fully sink in and all of the ramifications to fully unfold.

When I was fifteen, my father passed away. He had been sick for six months and was not in the best of health. It was a running joke that it was "good news" I didn't have his genes because of all the bad heredity he and his side of the family had. My father is and will always be the man that raised me those first fifteen years.

Five years later, I am expecting a son of my own. The doctor sits us down to talk about our family history. I explain my mother's side and then trail off.... This has always been a difficult topic. Whenever anyone starts about their ethnic makeup, I always tell them what I know from mother's side of the family. I have often looked at my son, who is now seven, and wondered if he might look like his biological grandfather at all.

My mother and I have talked about this only two or three times since that first time. She said she wouldn't have a problem with my trying to find out more information about my donor father. There's a sense that it should be easier because my father has passed on and there would be no hard feelings or misgivings. But I've realized that I carry a lot of misgivings about myself and my identity, specifically my biological and ethnic history. I don't think I'd even want to meet the man, just to see a medical chart, genealogy, and maybe a picture. But even after sixteen years of knowing the truth, I'm still not sure; I guess I still feel conflicted about the whole thing.

"Donor-conceived Child Who Is Now Able to Trace the Man Who Is Supposedly My 'Father' "

I have been old enough to find out who my biological father is for a few years now—although I still just think of him as the sperm donor my mum chose. She wanted a known donor so I could find out about him if I wanted to—and she talked about her yearning to have a child but that she didn't want to stay in a relationship just to have me. So she used a known donor. When I was one, she met my dad—whom she married—and although he isn't my dad by DNA, he raised me. He is my dad. They divorced when I was thirteen, but he is still my dad.

When I got to eighteen, I thought perhaps I would want to find out more—just for interest's sake—but I don't. Perhaps that will change. What matters is that I have family—even if not all are blood related. I am so loved. They were honest with me. My real dad couldn't love me more. It's all I need.

The man who is my biological father could be a horrid man—and once I know, I know. Of course he might not—but I don't need another dad. I have one.

I'm glad she chose a known donor so I had that choice. But when it came to it, I don't need or want to know—and, after all, he was just a stranger who donated his sperm. Anyone can do that. It takes so much more to actually give a child what they need—and what I needed was just what I got.

This story is in case my parents ever see this site and think somehow I resent it—or need to know my biological father. I don't. I'm not hiding my true feelings from you—we've talked about them since I was a baby!

I love you, mum and dad. Thank you so much for bringing me into this world. I am blessed.

"The Dream"

Regrets and Redemption—Truth from the Hereafter

I am in a white space. On one side of me, all the people in my life who have passed away, who I loved and grieve for, forming a half circle around me, radiating pure love and acceptance. My mom, my dad, my mother's mother and father, my dad's mother and father, and my stepfather. But the circle isn't complete. The other side of the circle is still empty. We are waiting for someone.

Then, a man walks towards me out of the mist. I know that he is my father. He's smiling and radiating the same pure love and acceptance as my known family. We hug for the first time. I feel euphoric. My father's mother and father follow, then their parents. The circle closes around me. We all embrace. No one's love or relationship or rights trumps another. I am finally complete, and I feel, for the first time, a sense of calm. All the grief, at last, is gone.

"Thankful to You"

[This is a response to "Silence," found in this volume on page 93.]

I have known that I am the child/product of sperm donation/whatever you want to call it, for six years. I am also in a graduate program (MSW) where I have to complete assignments—such as genograms—and think about issues constantly. In my undergraduate program, I had to read a book about sperm donation (for a family law class) and write a ten-page paper on the book and the topic. Sperm donation is everywhere. People make inconsiderate jokes, not having a clue that someone they know is dealing with the very same issue. Before I told my friends, one of them made a comment when a commercial came on the radio about egg donation, saying, "Why the hell would anyone do this to their kid?" This was my best friend. I sat in the bathroom and cried for an hour, but she had no idea. I got asked on a date to see Jennifer Aniston's movie. Even this past weekend, a friend of mine, totally unknowing of my situation, started talking about sperm and egg donation. It's a hot topic and people have opinions on it. People also *love* to ask the question, "What are you?" in reference to heritage. Coming up with an answer, a lie, on the spot, is never fun. Even worse is when you get caught in the lies. It's impossible to escape. The reminders are everywhere.

I have spent six years looking for an outlet. I have the most wonderful friends, but whenever I broach the topic with them, they get awkward, because they do not have the words to say to me. My family wants to discuss nothing about it. No one understands. Reading what you wrote, I have hairs standing up on my arms. I could not have phrased what you said better myself, and I am a lady of many words. After reading your post, I *honestly* for the first time have not felt alone about this issue. Thank you for sharing your story, because through your sharing, you have given me some solace and peace of mind that I have not yet experienced surrounding this issue.

My point in writing this is to emphasize how *beneficial* these stories are, at least to me. To my knowledge, there aren't support groups around for people dealing with these issues. This website seems to be the next best thing.

Thankful.

"Completely Shocked!"

I am twenty-nine years old and am facing fertility issues. I had honestly thought about if I might ever need donated sperm for my husband and I to get pregnant. After my first insemination with my husband's sperm, my mom sat me down and told me that my conception was the result of insemination via donor sperm. My dad isn't really my biological dad.

I grew up thinking I was adopted at times. I do not look much like my brother (who is biologically both my mother and father's), and I don't look much like my parents either (a bit like my mom). I grew up in constant fear of many diseases which run on my dad's side of the family, and in fact, went through countless medical tests growing up to be tested for these (and yet my parents knew that my chances of these would be slim).

I work in the science/healthcare field. I believe in genetics. It's hard to not look into the mirror and wonder where certain features come from, or if medical problems that I have been experiencing are related to my biological father's side. I am concerned about what surprise medical problems I could face, as when my mother received the sperm (it isn't like now, when women can choose many different qualities about the donor or medical histories), all she was told was hair color, eye color, and ethnicity. She accidentally ran into the supposed donor in the waiting room of her clinic (yes, this was before sperm banks).

I have known this information for two months now, and I still have difficulty processing it. I would really like to be able to receive health information back from the donor, but I don't even know where to begin. I know that I would *not* use donor sperm personally, because I now know the difficulties of learning this information. I love my dad dearly and wouldn't change anything about my life, but would still like to know the roots of my biological father. I wish there was more support out there for those facing this kind of life-altering information. I am not angry at my parents, but still struggle with the secrecy, as I have not told my brother that we are only half-siblings, and my dad doesn't know that I have been told this information. I was treated differently by his parents growing up, and I now know why this occurred (they were aware of the donation).

"Jaws of Life"

I found out my biological father was a vial of frozen sperm labelled "C11" when I was twenty-one. Finding out so late was a huge shock. With my childhood already behind me, the neural connections identifying my dad as my dad were cemented. Emotionally I could never think of him as anything other than my dad (and I still don't), yet suddenly I was told we were genetic strangers. My identity had been splintered, and the social and biological aspects of parenthood carved up. In the place where I inherited half my genes, all I could see was a vial of semen in cold storage. I mourned the human face behind that vial, somebody I had never and would never meet. A little bit like a mother might mourn the baby she could never have, I suppose.

I wonder what it would feel like to have been told earlier. Another donor-conceived man I met has always known his origins. Initially he accepted it. He even considered becoming a donor himself. On the day his daughter was born, as a father, he glimpsed the power of the biological link and what the loss of his paternal kin meant for him (and his daughter). I suppose the point I am making is that children do not have a static response to being donor conceived; it changes throughout their lives.

I couldn't relate to my story. I am a human being, yet I was conceived with a technique that had its origins in animal husbandry. Worst of all, farmers kept better records of their cattle's genealogy than assisted reproductive clinics had kept for the donor-conceived people of my era. It also made me feel strange to think that my genes were spliced together from two people who were never in love, never danced together, had never even met one another.

My reaction was never to blame my parents. I wasn't angry with them. In some ways I felt like my mother was a victim of telling me the truth, that she needed me to comfort her and tell her that it was ok, I didn't need to know who my donor was. I struggled to find the words to express my thoughts. The questions that I dared not ask, or even form in my head, because it seemed like a betrayal of my loyalty to my family and to society.

At the time these thoughts were incoherent, but I believe they basically boiled down to this: "How could my own parents decide to deliberately separate me from my kin, to grow up half blinded to my own identity? If they couldn't face telling me the truth about what they had done, why did they do it?"

"How could the doctors, sworn to 'first do no harm' create the system where I now face the pain and loss of my own identity and heritage?"

"How could the government, charged with protecting the most vulnerable members of the community, its children, legislate to make it illegal for me to know the identity of my biological father? How can its institutions subject me to the psychological torture of knowing that records exist but I am forbidden to know the contents?"

"How could my donor help create me, and then abandon me without even leaving his name?"

The best I have come up with to answer these questions is

"My parents were focused on the immediacy of their own infertility and would have done almost anything to relieve their suffering and get a baby."

"The doctors were focused on publishing their next scientific journal paper, and were surrounded by images of smiling, happy babies on their clinic walls. They didn't think about the future, when these babies would grow up."

"The government found it messy and awkward to legislate in this area, and there were no votes in it."

"My donor was young and focused on doing 'a good deed.' He believed the clinics who told him that the biological link can be extinguished by signing a contract."

For three years, I hardly talked about these confusing thoughts with anybody. Our family life continued pretty much as it was before. I reached a turning point when I met another donor-conceived woman. It was a huge relief to talk to people with a similar background, who shared my view that donor-conceived people have a basic right to information about their genetic identity. These people helped me articulate the things that were bothering me, and I bonded with an adopted woman who had seen it all before with the adoption debate and eventual reforms. Being donor conceived is like being "half adopted" but with the added strangeness of being raised by a blend of both the adopted and birth family.

I became driven by a sense of injustice. I had two newspaper articles published in the newspaper and learned the power of personal stories. I stood up in a seminar and asked the attorney general what he was doing for donor-conceived people. I faked confidence. I convinced a lawyer to engage me pro bono. I met politicians deep in the bowels of Parliament House and traded on my personal story to try to pierce their rhinoceros-thick skin. When the question of unsealing all donor-conception records was put to a vote, the amendment was defeated by a measly five votes. Five strangers had decided my fate was to have my questions go unanswered forever.

Or had they? I got in touch with my mother's treating doctor and asked him to send a letter to my donor, on my behalf, asking his consent to exchange of information and/or contact. The doctor was a highly decorated expert in donor conception. I was the first donor-conceived person he had ever met.

Three agonising months later, he emailed me to say he had done it. After this, things moved really quickly. The very next week, I received a call with big news. Firstly, I didn't need to ever refer to my biological father as my donor. His name is Chad.

After exchanging letters and talking on the phone, we arranged to meet. The day before, I discovered I would be meeting his children, my half-siblings. I was nervous, especially the night before and day of the meeting. As I approached the gate, Chad's son called out to me in an excited voice and ran to greet me. Immediately I felt more at ease. I said hello to everyone and we sat down to lunch. I had a surreal moment as I looked around and realized I was surrounded by people who all looked like me.

The clinics were wrong. We are family, at least in some sense of the word.

Finally I understand why people comment that my sister looks Swedish and why I am interested in flying and space. Chad and I share an interest in reading, art, sports, napping, nature, and the outdoors. After all my efforts in the media, law, and political lobbying, I was pleased to discover that my paternal grandfather was a notorious agitator of the establishment.

A few weeks later, I met Chad's eldest daughter. She gave me a card that read, "Dear Becky, what a wonderful surprise it was to learn about your presence. One can never have enough family, lots of love, xoxo."

For me, the hardest things about being donor conceived was the powerlessness and lack of choice—being constantly reminded that I must abide by decisions made long ago. Hang on a minute; I never agreed to any of this!

The other hardest thing was seeing how society had accepted and valued the biological link in endorsing my mother's need to have a child she was biologically related to, but rejected, sometimes ridiculed, and at the very least, constantly required me to justify why I needed to mend the broken ties of, my biological link to Chad and his family.

"How My Mother Told Me"

When I was thirteen years old, my mother told me, "Daddy's not your real father." I do not remember what started this conversation. We were sitting on the couch in the living room, and I think we might have been talking about the way Alzheimer's runs in my dad's side of the family, because next she said, "I thought you'd be relieved." She sounded irritated when she said it, probably in reaction to the way my jaw had dropped open in shock. (Until that moment, I had always assumed the jaw dropping open in shock was just something that was done in movies and cartoons to convey surprise, but it really happens. It felt like a hinge had broken, and I had to use my hand to physically close my mouth.) Maybe it was apparent I was going to cry. I cried a lot back then. My mother explained, as my dad sat in the next room oblivious to our conversation, that Daddy wasn't my real father, because she had been inseminated with sperm from an anonymous donor at her doctor's office. The donor was a med student, she said, though I wonder now if this is something the doctor actually told her or if she decided this fact on her own after hearing that donors were often med students. The only other things she knew about him were that he had the same hair and eye color as my dad so that I would convincingly be able to "pass" as his daughter.

"I'm surprised you didn't figure it out on your own," she said. "You're so much better at math than me or Daddy. How else did you think you got to be that way?" My mother told me not to start crying. She really hated when I cried, and I suspect if she had foreseen my reaction, she never would have told me about my father. There wasn't a door between the living room and where my dad sat in the family room, and she warned me that I was not to tell anyone this secret ever, particularly my dad. He knew, of course, but my mom said he seemed to have forgotten, and aside from her own mother and that doctor whose name I never learned, no one else in the world knew I wasn't my dad's biological daughter, and they shouldn't. Besides, my dad hadn't even wanted me when I was conceived, my mom said. He'd told her he hoped she miscarried.

My mother did not seem to understand why any of this news was upsetting to me. After all, she said, knowing that my dad wasn't my biological father didn't "change anything." My biological father wasn't my father anyway—he was just a "donor." It's different. "It was anonymous," she said, "so you can't ever know who he is. We agreed never to find out, and if you ever try, I'm the one who will get in trouble." A note to anyone who has a donor-conceived child: This is not the best way to tell them.

My mother said the only way to unlock the donor records was if I were to get a terminal illness, since having my father's medical information could help the doctors treat me. It turns out she was wrong about that—there was no way to find out the identity of my "donor" under any circumstances. According to the office where my mother was inseminated, her records had already been destroyed by the time she told me this story. But I don't think either of us knew that at the time. A fact I've never told anyone else until now is that I prayed for a disease that would allow me to unlock the file that I imagined would tell me who my father was and what siblings I had. I prayed for it every night for I'm not sure how long—at least a few weeks, but it seemed like a few months—as I cried myself to sleep. If this seems stupid or petty, please consider that I was thirteen at the time and that this might have seemed a more pressing issue to me than my own mortality.

"Don't tell anyone," my mother repeated. "If you really need to talk to someone, we'll get you a therapist." And that was the end of the conversation. After the first couple weeks of crying myself to sleep, I asked her to set up the appointment. I'd always been a crier and good at keeping secrets, but I was having an abnormally hard time coping with this news on my own. My mother then told me therapy would be a waste of money, like it had been when they'd taken my brother to a psychologist, and all I'd do is blame her for everything anyway—that's what therapists get you to do, she said. So I kept my secret entirely to myself, not mentioning it again even once for the next three years.

I am twenty-nine years old now. I am married. I have a house and a dog and a college degree. I didn't get a terminal illness, but I did get a chronic one my doctor says must run in my father's family. But it's not that bad, and I don't blame the donor for not knowing about it. I never again mentioned my biological father to my mom or anyone in our family. I am closer to my dad now than I used to be but more distant from my mother.

My desire to know who my biological father is has not really diminished in the years since I learned of his existence. I don't particularly like him, since I feel he gave me the ultimate blow-off when he agreed to sire me in exchange for money and a promise never to find out who I am or even how many of us exist, and he accepted this arrangement as a good deal. I don't particularly value his decision-making skills for the same reason. I don't want back child-support payments, as some donors fear. I don't want his love or to call him "Dad"—I already have a dad. I don't want to be on his family's Christmas cards or to take up an inordinate amount of his time. I just want to know who he is. I want to know if I have siblings and who they are. I would like to have lunch with him and make polite conversation with him in which we call each other by our first names and talk about what we do for a living or where we like to vacation. Or something. I'd like to talk with him about **something**.

Some proponents of anonymous donation say that plenty of children grow up not knowing where they come from because they are put up for adoption or because their fathers ran out on their mothers, and anonymous donation should be allowed because it isn't any worse. My thought is that none of those situations sound great. If we can correct one of them, why shouldn't we? I should add that I am in the U.S., where there are currently no laws granting an eventual right to information to donor-conceived offspring.

I spent many years feeling ashamed and ungrateful for wanting to know who my biological father was. I don't anymore. If studying genealogy and mapping generations of history onto your family tree are reasonable hobbies, I am allowed to be interested in who my father is, even if there is no way I will ever know him. It is okay for me to want to know, and from what I have read by other donor-conceived offspring, it is staggeringly normal. No amount of semantics or reasoning or wishful thinking on the part of parents and donors will be able to change that.

"Missing Pieces"

I still remember finding out as if it was yesterday. A few days after my thirteenth birthday, I found out my social father had myeloma (bone-marrow cancer). About a week after finding out, he had a reaction to the steroids he'd been given, and the only way I can put this is, he went a bit funny in the head. Maybe a "bit" is an understatement.

It started like any normal day, although I can remember noticing my social father being slightly more grumpy than usual. About 3 p.m., my mum came into my room in tears and said that they needed to talk to me. Mum led me into the lounge and they sat either side of me on the sofa. What always stays prominent in my mind when I look back on that day is the first thing my social dad said…"This is the last cry for help." I still don't know exactly what he meant by that (I don't think my mum did either), but it was the way he said it, the way he kept repeating it. I knew something was really up.

Then they told me I was conceived by a sperm donor. That bit was all a blur. I remember being told, not saying anything, and going up to my room and crying/sleeping for the rest of the day. I didn't come out until the late evening when mum eventually realised that my social dad really wasn't right and we had to call an ambulance. I remember following the ambulance, not saying anything to my Mum, getting to the hospital and seeing the paramedics who had brought my social dad in. They said "Ohhh we've heard a lot about you!!" and started to laugh. I followed them in silence to find my social dad in a bed in the hospital ward. One of the nurses said to me, "Why don't you give your dad a hug? He's had a pretty rough day." And I thought, "if only you knew…" but that was when something clicked inside me, he wasn't my dad, he never was, never will be. Yes he brought me up, paid for me, but we never shared a certain connection. I remember a few years before I was told, asking my mum if I was adopted because I always knew I wasn't like my social father. Later I found out that it was my social father who had been adamant to tell me, but only because of the "funny turn."

For the next few months, it was only spoken of if I had a question and wanted to ask my mum, and I had plenty of questions. I still do. The worst thing about it is the most important questions I want to ask, my mum can't answer, and I will more than likely never find the answers I'm looking for. To this day I feel like it is a taboo subject with my parents. They know that

I'd like to find that side of my family, but they never bring it up. In a way, I resent my parents for that. It's a part of me; in fact, it's half of me. Half of my genetics. They made the decision to bring me into the world that way; they should deal with the consequences of me finding out. Of course I want to feel free to talk about it and not have to worry about causing upset. They'd never told anyone until I found out and then gave me free reign to tell my Auntie and Uncle (who then told cousins, I think) and my Gran. Luckily my social father's family all live in Australia, so we've never seen the point in telling them. I'm so glad that my family knows now, but they all see it as though he is still my dad, I should still refer to him as Dad (which I put off doing as much as I possibly can because it makes me feel so awkward). I do love him, but not in the same sense as I love my Mum. We don't have the same bond.

I'm nineteen now, still in the process of registering with Donorlink UK. It still hurts to this day, not quite as much, but it still hurts. It makes me want to shout and scream at parents who are considering using donor conception—tell your children from a young age, answer all their questions, relate to them!!!! If my parents could see this website, then maybe they'd get an insight on how it feels to be me. But I have to be so careful not to upset anyone about it, when really, it's me that's upset! I didn't ask to be conceived this way, it was thrust upon me that I don't know half of my biological family. I want to know my medical history, perhaps what career my biological father has, how many half-siblings I have, and my cultural ties. I want to know this even more so for my own kids. I read an entry earlier on this website where someone said I couldn't lie to my own children about it, like my parents did to me. That's exactly how I feel; I don't want my children to grow up thinking they are biologically related to a grandfather that they aren't actually related to. That's not fair. He can play the part of granddad, they can call him Granddad but they are going to know that out there somewhere is their real granddad. I want to know that half of my medical history as well, for my future children's sake.

I genuinely feel that had I been told from a young age, and had grown up aware that he wasn't biologically related to me, that it wouldn't have had such a massive impact on my life. I don't think anyone of my friends/family quite realises how much it does impact my life. Don't get me wrong; I am unbelievably happy and relieved that I did get told. I just wish my parents had actually made the decision to tell me, rather than it coming out by accident.

I am very happy for future donor-conceived children that they have taken away anonymity for donors in the UK. But the people born before then don't get a say. If I had been born three months later than I was (before the August 1991 HFEA changes), I would have been entitled to receive nonidentifying information about my biological father when I reached eighteen. So what about the huge amount of people suffering who were born before then?

I would love to meet another donor-conceived person, as I have talked to a few online but never actually met anyone like me. Everyone's stories that I have read from this website I can relate to in some way or another, and it is so reassuring to know that I'm not being selfish/attention seeking and that others think exactly the way I do about being donor conceived. No one knows what it feels like until they are put in that position. They can say they "relate," but they can't. Everyone says to me, "But he's your dad." He's not. I'm still looking for him.

"Alone"

I am eighteen years old and have known I am donor conceived my entire life. I stumbled across this site by accident. I was intrigued by the stories. Many of them deeply upset me. It has always made me feel bad that I will never know the identity of my biological father. I think of how precious it is to know where you come from and why you look the way you do.

The holidays always make me think about it even more. I wonder where he is, what his traditions are, where he is and what he is doing. I hope he thinks of me (the potential offspring). I wish he simply knew I exist. It is the night before Thanksgiving, and it really has me thinking of all this. I *tried* to share with my mother how I feel. She did not want anything to do with it. She makes me feel as though I should not have any feelings on the subject of being donor conceived. She especially gets angry because I do this every year around this time since I entered high school. I really feel like I cannot help this. She never will understand how I feel. I wish she would understand. I don't want her to blame herself for my misery. I try not to broach the subject at all with my nonbiological father. I love my parents; really, I do. I also know they love me. I sometimes wonder if they are putting on an act. I wish I could go deep within their minds. I would love to know how they really feel about the subject.

I am a young woman who is asked to keep quiet about her innermost conflict. I entered college in the fall and have had an emotional ride coming to terms with being a donor-conceived person. I need a place to vent my feelings. I just wish I could find someone who would understand. I feel like I am suffering alone with this. I don't want to disappoint anyone. If I could have one wish it would be to just hear the donor's voice. I would love to ask him the most basic questions about himself. I wonder why he did it. It bothers me that there was money involved in my conception. I resent the donor for that. I feel as though he took advantage of my parents and the parents of my countless half siblings. If he did this for the right reason, he would have done it for free. I also resent him for not taking advantage of the Donor Sibling Registry. I find it cruel that he does not want to find me or any other offspring. I hope one day it will all be all right with me. I still hope that maybe I will meet or talk to the donor. I have had the opportunity to meet one of my half siblings. I was happy to have gained some insight into who the donor is through my sibling.

"Beginnings and Ends—A Matter of Parental Choice"

One of my earliest memories is of lying in my pram.

Few people would believe that, as many adults have difficulty remembering their first day at school, let alone anything that happened when they were only toddlers, but remember it I do!

I was probably a little over a year old at the time and lying down in my pram because I was strapped firmly in place by my harness. I was supposed to be taking an afternoon nap while my mother was busy elsewhere, but I was bored and not at all tired. I wanted to be up and about, not a prisoner in my pram. The pram hood was up to shield me from the light coming in the lounge window and effectively all I could see from my prostrate position was the top edge of the closed door and an area of faded yellow wallpaper. I yelled for attention. I kept on shouting and screaming until my throat was hoarse, but still nobody came. Then finally, when I had all but given up hope of being released, I saw the door swing open, and I knew that my mother had heard me, so I expected her arms to appear around the side of the hood and for her to unclip me and lift me out. Instead, I saw my frilly white pram pillow appear in my line of vision. It was advancing towards me, taking up more and more of my view of the wallpaper until I could see nothing else. As it came closer still, it blocked out all the light and then it was right on top of my face, pressing down on me, and I couldn't breathe. Then there was blackness and swimming heat, muffled noises and nothingness, until the next thing I remember, which is being out of the pram and in the light again, hearing my grandmother's voice fussing over me in a reassuring way.

If my grandmother had not arrived unexpectedly that afternoon and interrupted my mother, I would have become a cot death statistic, but nobody will want to believe that my own mother had tried to kill me, especially as she had spent five years and a lot of money desperately trying to have me. The experts would be the first to argue that being donor conceived means that I must have been very much wanted and loved, so my memory of being temporarily snuffed out by my mother can only be the product of my own warped imagination. I might even be inclined to think so myself, were it not for discovering two other donor-conceived adults who have had similar brushes with death at the hands of their parents. In one case, a teenaged girl woke up to find herself locked in her bedroom after her mother had left the gas tap to her fire switched on, but not lit. She could have been dead by morning, but like me, she has survived to tell the tale. In another case, a social father who had trouble reconciling his infertility has admitted that his ambivalent feelings towards his baby son had led him to the brink of committing infanticide. The horrifying fact is that we three donor-conceived people have survived an attempted murder by a parent.

Donor conception is not a cure for clinical infertility. It doesn't give parents the exact baby they had hoped for, and consequently the replacement child may never live up to the happy-ever-after fantasy to which the parents had aspired. In the circumstances, it is inevitable that some parents will direct their bitterness and resentment at the child who is a constant reminder of their reproductive disappointment, particularly if the balance of their minds has been disturbed either by their infertility, their choice of circumvention, or a combination of the two. I was not at all surprised by a recent report, which showed that a surprisingly large number of healthy donor-conceived babies are aborted before birth because their mothers have changed their minds about the pregnancy. It is, after all, only one small step away from deciding to terminate the life of the child once it has been born. Do I support donor conception in any way, shape, or form? Certainly not!

"Safety in Biology"

When I tell people I wish I had been raised by my real father, often they are offended and say things like, "Your real father is the man that raises you, kisses your scrapes, tucks you in, reads you stories, and loves you." This is a wildly naive idea, and if donor conception and the adults that use it could reliably provide social fathers that do these types of things, then maybe I wouldn't have so many problems with the practice.

I strongly believe, because of his lack of biological connection, social fathers are generally much less likely to sufficiently raise their children. Let me say it another way: Nature hinders man's ability to raise a child that is not his biological offspring as well and with as much grace and love, as the child that *is* his biologically.

When my parents divorced, I was shocked at how easy it was for my "dad" to sign away custody rights for me. I haven't seen or spoken with him in two decades. This is the man that agreed to "raise me, kiss my scrapes, tuck me in, and read me stories." It was a lie. Then my mom remarried. Her new husband agreed to do these things for me also. I was happy to have someone around to horseplay with and lift me up. I liked to wrestle and do a lot of gymnastics as a kid. Soon after he adopted me, he started touching me inappropriately when I reached out to horseplay. When I told him to stop, he said I didn't know what I was talking about, and we were just having fun. I stopped horseplaying. I stopped doing gymnastics around the house. I stopped feeling free in my body. I stopped being a kid.

As I started to develop, I tried to stay far away from him. I tried not to engage in conversation with him. I tried to be polite, but I was incredibly creeped out by him constantly. It seemed he always really needed my attention when I was in the middle of taking a shower. I'd be changing, getting the water running, and he would always open the bathroom door and have something super urgent to show me, like some BS news story. I knew he was getting a peek at my body. I started locking the door. And I started hating him.

I asked my mom if I could be transferred to a boarding school so I wouldn't have to live with him. She had no idea why I would want to be away from her. She loved me and needed me so much she didn't want me to go. She had no idea about the touching and didn't seem to notice the sexual harassment. I didn't have a word for the sexual harassment; I just endured it, because there was nowhere safe for me to go.

I've tried to bring up sexual abuse as it relates to donor conception before, and some say it has nothing to do with the practice. Children are *eight times* as likely to be abused by a stepfather or "social" father. For the ones that are good dads, I'm really happy for them, and keep up the good work, but artificial insemination for straight women inherently means that they are being raised by a man or men that have no biological link to them. Biology adds safety.

It doesn't matter how smart you think you are or your husband is. My stepdad had more degrees than he knew what to do with. This is something bigger than human intelligence—it's nature.

"I Have a Right to Know"

I am an adult woman. I am a professional. I have a master's degree. I have many friends, a large social and professional community, and so many things to be thankful for. I do not think that because things have worked out for me that I should not desire to know where I come from.

I don't doubt that my family loves me. I don't doubt that I was really wanted. I don't doubt all of the right intentions were present in my parents' hearts when they went through the process of creating me, with science.

I want to know why I'm different. I want to know if there is someone else who can instinctively know, know without words, who I am and have it make sense. There is a connection between people, which is more than just social, more than experiential, more than upbringing. There is something biological, and I can see it in others. And I can see how I lack it when I'm with my family.

Do I have grandparents? Do I have siblings who laugh like I do? Are there people in my community, like me, who maybe share a link which makes me makes sense? I'm glad to be alive. I'm glad to have the life I have. I'm sad that I know so little about where I come from.

Families look different for everyone. And most people can describe an undesirable element in their family. But it's their family. Where is mine? And why is there a system in place that prevents me from knowing? Knowing my lineage, my genetic background, my family, is a human right. It is my right.

"The Politics of Donor Conception"

As a donor-conceived person, I have the sense of being part of an underclass. Because I was born in the early 80s, before the changes in UK law, I have no right to know the identity of my biological father. I can't think of any other group of people in society who have this information denied to them. I feel like half a person. Even though donor-conceived people born in the UK after 2005 will be able to trace their donors when they turn eighteen, it can be argued that it is wrong to bring a child into the world knowing they will be unable to have a relationship with one of their biological parents (which the child may or may not desire, and why shouldn't they?) until they reach adulthood. At best, donor conception is a poor way to have a child.

The changes in the law will allow donor-conceived people to trace their donors only if they know the truth about their origins. Unfortunately, though parents are encouraged to tell their offspring, the law does not require donor conception to be recorded (e.g. on birth certificates), so many children will still be kept in the dark and may experience the same sense of being somehow less than human that I have done if and when they eventually find out. My mum and dad never told anybody, but the truth still finally came to light. (I found out by accident when I was pregnant with my son and a routine blood test showed that I have blood group AB—because my dad is blood group O, I knew he couldn't be my biological father.)

Having a child is a privilege not a right, and I say this as a mother of two children—one who I lost through miscarriage and one who is now a healthy toddler—who would dearly like more. Potential parents considering donor conception would do well to bear this in mind.

It is my firm belief that we all have a human right to know the truth about how we came to exist and the identities of our parents. As far as I'm concerned, I have three parents, equal in importance: my mum, my dad, and my biological father. I wouldn't be the person I am without the input of all three.

"Thank You, Ben and Jerry!"

I was twenty years old, home from college for the weekend. I was sharing a pint of Ben & Jerry's ice cream with my mother. We were watching some sappy made-for-television movie about donor conception/adoption. I can't remember the details. I said to my mother, "If I had a problem conceiving a child, I would definitely consider doing it." To that, my mother replied matter-of-factly, "How do you think you got here?"

I was shocked and yet gleefully happy to learn that her husband was not my biological father. After all, we had nothing in common and I detested him ever since I can remember. But what was I to do? Where could I turn for help? My mother wanted me to keep it quiet so that no one else would find out. She didn't want my father's mother to know in case she would write me out of her will! She didn't want me to tell my younger sister (also donor conceived and looks nothing like any of us) for fear it might upset her.

There are no records to find. Mom's doctor is deceased. I've checked Donor Sibling boards and found a possible donor's name but cannot locate him.

I'm a college-educated professional. I want nothing from my bio-dad except to see him and talk to him. I'm insanely curious. What does he look like, how does he sound, what would it be like to share a warm hug? I've never had that kind of relationship with my "mother's husband." It would be so nice to know that my real father is no longer absent.

"Dear Anonymous Sperm Donor"

This is a letter to a now deceased anonymous sperm donor. I wish I had been able to meet you while we both were alive. I do believe in an afterlife, so maybe we will meet up somewhere in the future. I would love to have known if my sense of humor was shared by you, what were your political opinions, if you were religious, how you laughed or cried or if you had a temper. I did have a dad who I do believe loved me, but we were so very different in almost all respects. He was reserved, conservative, very religious, had no sense of humor, and was not what you would call warm and inviting. He and I disagreed, and if I made him angry (which I suppose I did, a lot) my mother would say to me, "You don't know what a wonderful father you have." I heard this quite a few times growing up. The reason I "did not know what a wonderful father I had" was because no one ever came right out and told me I was donor conceived and that my father had allowed my mother to have a child by this method.

It took me years to put all the pieces together. From the night I heard my grandfather tell a dinner guest, when I was nine, that my parents had only one child because my father had had the mumps when he was an adult and my mother had to go to a special doctor in Chicago in order to get pregnant. Now even at nine, I could figure out that if my father had the mumps, it didn't make much sense for my mother to have to go to a special doctor. So I saved all this info, and as a very inquisitive child, I added to it over the years. From why my mother and father had very, very blue eyes and I had brown eyes, (We learned this was pretty much impossible in high-school biology) to lots of other little things that just didn't add up.

So, anonymous father, who were you? Well, as I said you are now dead (unless you have very good genes, one can only hope), and I am sixty-seven years old and am still trying to find you or who you were. I just spent $300 on DNA testing that could possibly unearth a cousin who might be your kin. I keep thinking that maybe there is a box somewhere with info in it that will produce your name or picture.

Maybe my mother, lying in a nursing home with last-stage dementia will suddenly wake up and say, "I just wanted to tell you who your biological father was." Maybe she doesn't know; maybe she just went to the doctor's office, layed down on the table, and had your sperm injected into her womb. Maybe the doctor went into the side room and contributed his own sperm. Maybe his son, who kindly agreed to do a DNA test with me four years ago (which turned out negative) and which I paid $250 to have analyzed, really is my half brother and the test was just not reliable. Maybe you had prosopagnosia like I do, and we could have talked about what a pain it is to not recognize people's faces after you have met them about twenty times. Maybe you died from melanoma because you were not as careful about looking at all the little spots on your body as I am and could have told me to be careful of my kids out in the sun when they were little.

Maybe you would have liked to meet your grandchildren. One is a VP in a company, has two kids of his own and a very lovely wife—he wouldn't be looking for anything from you but would have found it fun to meet his biological grandfather. The other is a successful legislative specialist, has two great kids and another super wife, but one of his kids had to have brain surgery for birth trauma, and they all could have used all the support they could get—maybe you or your family could have been there for them, even if it was just by saying a prayer. So, you see, biology is important. Otherwise why would I spend time, money, and energy searching for you. I hope you had a good, happy, and long life. I am sorry I missed you; maybe our paths will cross again.

"Filling Out the Story"

"Where are you from?" they ask (all the time). So you tell them where you grew up. "No, no.…I mean where are you *from*? What is your ancestry?" This is so American. People look at your face, and they always want to know who blended with whom from where to create your face. When I was little, it wasn't such a big deal. My mom spoke mostly for me, and when I answered on my own, it was all regurgitation. As I grew older and began answering for myself entirely, spending the night at friends' houses, and meeting people on my own, it became harder and harder to give clear answers. "Mixed" doesn't satisfy anyone, including myself.

I knew absolutely nothing of my father's history. I knew nothing of half of my ancestry. I managed to find out a little bit of information on him, but it took forever and wasn't much. Those few extra details began to form a shell of a character though. That's when I really started to imagine him as my father and fall in love with the idea of meeting him and knowing him. Those few details can inspire a whole person, and I felt robbed of him.

After I found those details about him, people would ask me about my father and I would tell them he died when I was a baby but he was from this country and did that and that and that for hobbies and work. I felt entitled to my father in the shell of a character he was. But that too would be traumatic, because people would ask how he met my mother or other questions I had no answers for, and lying became improvisational and stressful as I tried to hide and cloud the fact that I'm donor conceived.

I hate being donor conceived. I think it is ridiculous and bizarre that the two people that made me have never met and never will meet. I think it's creepy that my dad was paid. I think it's creepier that agents and salespeople and commercial doctors worked so hard to create me and now that I'm an adult have no interest in my opinion. They're like drug pushers.

Selling substances that cure baby cravings. Do they have anything to cure my father cravings?

"Dear Fertility Industry"

Dear Fertility Industry,

You ruined my relationship with my mother. For twenty-two years of my life, she told me everything and was my best friend. But *you* told her that it would screw me up if I knew that "my dad wasn't my dad"—that she should keep it a secret. You don't think finding out from someone else at twenty-two would screw things up a bit?! Why wouldn't you think of the consequences of what you were doing—why wouldn't you give any long-term thought to something so important?

Years ago I talked to my mom every day on the phone without fail. Now we're lucky to get ten minutes in once a week. Despite our best efforts, our relationship has never been the same since I found everything out. The secrecy, deception, and ignorance surrounding human reproduction are the reasons why.

Love, Disgruntled Sperm-donor Baby

"Silence"

I am in graduate school, and just this past week in a class, we were told our first assignment was a genogram, tracking our biological family. My heart sank. I can't go anywhere without being reminded that I will never know my biological father. It is on the news, it's what Jennifer Aniston is talking about, it's in a preview that is on every five minutes. It isn't just the media either—more often than I still can even comprehend, my friends, coworkers, anyone will proclaim "I look just like my dad" or some other biological-family tidbit. One of my least favorite things is sperm-donor jokes, which are also inexplicably common. The only thing I hate more than sperm-donor jokes is when someone calls their absent dad/child's father a sperm donor. Example: "We don't refer to him by name in our house. We refer to him as The Sperm Donor." I literally have to walk away in order to not get incredibly angry at whoever is saying it. In some way, shape, or form, there is always something that will remind me of being donor conceived on a daily basis. It never ends.

I feel silenced, and I have felt silenced since I found out I was conceived by an anonymous sperm donor when I was sixteen (the nine-year anniversary of finding out is in February). I have always known that something was "off" in terms of my family, but I knew I wasn't adopted—I look a lot like my mother and saw pictures of us in the hospital when I was just born. I had asked questions of my parents about if my dad was my "real dad"—a phrase I cringe at now—and they always assured me that he was. They lied, and I understand why they lied; I was young and they wanted to protect me. I have come to peace with the fact that they lied.

What I have not come to peace with is how silenced I feel. I feel silenced for so many reasons: because it's weird and people don't get it, because I don't want to hurt my dad—whom I absolutely adore—talking about it, and also because I am always afraid I will burst into tears talking about it, no matter who is listening. I also know that my parents are very private people in general, and to have a bunch of people—whoever they may be—know that I am donor conceived would make them uncomfortable. Despite this, they have told me that it is my story to tell and they are accepting no matter what. But I still feel silenced on the topic.

I have felt differently about being donor conceived throughout the nine years that I have known. I go through phases where I think about it more, or less, and phases where it weighs on me more heavily, or not. The past year has been the longest period I have thought about it quite regularly and frequently, and this past year, it has weighed on me the most since the year after I found out. I think a part of that is that now that I am almost twenty-five and my older friends are starting to get pregnant and have babies, which leads me to think about my own children I hope to have some day.

This presents a real problem for me. I don't want to lie to my kids about how I was conceived, the way that my parents lied to me. I also don't want to hurt my dad or confuse my children or somehow influence how they feel about their grandfather with that revelation. I try to tell myself that it doesn't matter as much, because it isn't (God willing) how they themselves were conceived, so it is a bit more removed. But deep down, I know that while that might be true, I also don't want to lie to them—or even withhold the truth. In addition to the sadness and discomfort that this presents, it also makes me angry that I even have to deal with the way I was conceived and that it will seep into my life for the rest of my life. What also seeps in is the feeling of being silenced and of feeling damned if I do, damned if I don't in many respects regarding being donor conceived.

I truly hope that they end anonymous sperm and egg donation. I don't think it should be banned altogether, but I do think anonymous donation should be, for so many reasons—some that can't even be identified with words. Again, the silence appears.

"Lies"

My parents never told me that I was conceived using a donor until I was well into my thirties with my own family. As bad as knowing that I may never find my biological father is, being lied to is the worst part, because now I wonder what else they've been lying about. And even when they say this is all to the story, there is still doubt in my mind.

13. Personal Stories from Donors, Surrogates, and Birthparents

"Don't"

I grew up in foster care and aged out at eighteen without ever having a family. I was a good kid, talented in the arts and academics. I worked hard with the hope that a family would want me. I was never adopted, so needless to say, I've been desperate to be a part of a family my entire life.

I was never paid for my eggs. I gave my eggs away. A few years ago, I gave my eggs to good friends with the naïve belief that this was a way for me to be part of a family—and I would even have a genetic connection! I was promised that I would always be part of the family and the child's life. I was told that I was always meant to be part of their family and that their family never felt complete until I came along. I wanted to give my friends what I never had—a family. I thought, "When would I ever have a chance to make such a huge difference in someone's life again?" And I would be part of a family too. How could I not do it? I was so eager for love and had no one looking out for my best interests, which made me easily exploited, even if unintentionally.

I took Follistim, Menopur, and Lupron. The doctors assured me there were no long-term side effects from egg donation. The only risk was OHSS, which only happened in 1 percent, they said. Knowing what it's like to yearn for family, I went through with the egg donation for my friends. I had to take some genetic tests for diseases, health tests, and a counseling session over the phone. The counseling session was about my life history and how the egg donation might affect my friends. It was never about my psychological needs. I produced an unreal forty-seven eggs. At first I felt pride in that. Now that I know better, I am angry that the doctors risked my health and allowed twice the normal amount of eggs normally produced by donors to develop and allowed my ovaries to stretch beyond safe levels. I was so uncomfortable. I looked pregnant. I had to hold up my distended abdomen just to walk during the last few days before the retrieval.

During the egg-donation process, the doctors and nurses were super supportive, except for when I questioned anything, like the fact that all the drugs came in my friend's name. When I questioned it, I was told, "Because she's the one paying for everything" in a pretty harsh tone. Um, she may have been paying financially, but she was not the only one paying for the donation. Plus, it's illegal to take drug prescriptions written for someone else, isn't it? After the egg retrieval, their attitude towards me changed. They did not want anything more to do with me, especially when I had complications. I became an annoyance. During the whole process, I had some eye problems. My vision was a little blurry, and my eyes' ability to focus changed. My eyes returned to normal about six months later, but when I asked the fertility doctor about it, he told me that he never heard of that happening. I later researched it and discovered that the FDA lists it as a side effect of the drugs I was given.

Immediately after the retrieval, I had a lot of bleeding and discomfort. I became severely depressed, to the point where I needed antidepressants but my mental health continued to decline. My periods stopped. My hair became dry and brittle. My skin changed, and I got a lot of acne. I gained weight at an alarming rate, but my diet did not change. Driving over bumps, jumping, or any type of rough movement was a little painful for months after my donation. I kept calling the clinic with my concerns, but each time I was told my side effects were normal and would get better. Six months without a period was normal? The side effects did not get better.

Finally I went to see my doctor, who ran tests and found that my hormone levels were "all over the place." He referred me to a specialist, who told me I had what seemed like PCOS, and then I was referred to an endocrinologist. I will be on drugs to control my hormones for the rest of my life. I still go months without a period. My mental health continues to be an issue and is very much linked to my hormones. My ovaries have remained enlarged, and I will most likely never be able to conceive naturally, if at all.

My friends promised me I could have the remaining embryos, but they have since changed their minds. They've changed their minds about a lot of things.

My friends have decided that I am no longer part of their family. She cannot handle that her son looks exactly like me. She is devastated that her son has no genetic link to her. She wishes he had come out looking like her husband, because I remind her that he's not hers. I'm not sure what she expected when she took my DNA. I'm no longer part of my "nephew's" life, after loving him for over two years; I have to stalk the father's blog to watch him grow up. I would have never given them my eggs had I thought I would not be a part of the child's life.

Despite my gift and sacrifices for the people I love, they have abandoned me. My devastation and drastic mood swings became too much for me, and I attempted to end my life. This was a scary blessing. Antidepressants didn't work well for me, so the psychiatrists ran tests and found the right medications to balance my hormone levels, and I feel so much better and a lot more stable. I donated my eggs out of a desire to mend the broken hearts of people that I love—and my own. What I got in return was a hormonal disorder, major depression, infertility, and another family abandoning me for reasons I do not understand. I wish I could go back in time and spare my body, fertility, mind, and my heart.

Egg donation is *not* harmless, like I was led to believe. It has seriously affected every part of my life. I often cry about the loss of my friends, my family, the little person with half my DNA, and my ability to have my own genetic children someday. If you're thinking about donating your eggs for *any* reason—DON'T. It's not worth it.

"Proud to Have Donated"

I became a sperm donor because I wanted to, not to make money. In the UK, where I live, there isn't the same level of financial compensation for donors as there is in other countries. I received £250 (around $385 US) for just under thirty donations. My travelling expenses were also reimbursed. Since 2005, the law in the UK has changed, and donors have had the right to anonymity removed.

I recognise the torment caused to donor-conceived children who do not have the right to know their biological father; in my view, the change in the law is a very positive one. In the UK, the number of families that can be treated with donor sperm is limited to ten. When my donor children reach eighteen, they will be able to trace me. I sincerely hope that they make this choice; I will be in my late fifties by this time, and I will probably welcome the company of younger people. Not all donors or potential donors feel the same, and the number of men coming forward has dropped considerably since the right to anonymity was removed.

My reasons for donating were very complex. Initially I wasn't that serious about it and just wanted to have the fertility test and sperm analysis. I did not realise that only 5 percent of men who come forward have sperm that is suitable for freezing. I realised that I almost had an obligation to continue with it. It was only after the initial tests that I really began to give serious thought to my reasons for doing it. Partly I wanted to help people who could not conceive; partly I wanted to send my genes forward into the future having no children of my own; partly it just seemed like something very positive to do. It is not very fashionable to be pro-human these days, but I really felt that adding to the population was a good thing to do. I felt that if a single woman or a couple were prepared to go to the lengths of being treated with donor sperm, then the resulting child would be wanted and loved.

I can understand that being the result of a medical process, sharing your genes with a stranger, having many half-siblings that are not known to you may be very difficult. I hope my donor children will be able to rationalise these difficulties, weighed against the fact that they have a life to live, a life that they would otherwise not have had, a life potentially filled with fascination and joy. Then I hope they won't appear to be such huge problems.

A Donor

"Kids Are Our Future! Proud to Be a Sperm Donor."

Since I was a little child I has always been concerned about the big questions about Life and Death and I also loved so much animals, so much that one day refused to eat the beef my mum was offering: "They killed this animal; I don't want to eat"

My mum was very concerned so she took my hands, she sat on the bed with me and showed me the decoration of the sheet: "Do you see this big flower? She is the cow and she is dead now, but can you see these little flowers starting from it? They are her babies. She is dead, but she will survive through them.…She won't be exctincted."

I think she told me this in an emotional peak, as after, I often thought about having a big family and my own children in future, and still I hope I will have a nice family to build up. :)

It was a long time ago, and now I am 21. I often heard my sister talking about some of her friends having troubles in conceiving, I heard about Cryos International, and since I wanted to have an Erasmus [student travel program] time, I thought, "Why not to donate? It would be nice to help other people."

I live in Italy, a country where sperm and egg donation is forbidden by law, so there are no sperm banks here, and that is bad.…But reading on the internet, I knew the existence of private sperm donation, so I put announcements last January and started having contacts with infertile couples, both Italian and English, and started meeting for home artificial inseminations.…

The first months nothing seemed to work, but then arrived July, the "lucky month," and two women fall pregnant, and now they are happily waiting their babies.

I can't describe the happiness of hearing the woman being so surprised and happy after years of sadness and desperation and yes....It was a very good feeling to know there is a new heart beating out there and that it is your biological descendent.

I feel happy in knowing my actions helped new lives to start; I live in Italy, one of the less prolific country of Europe....There are only 1.3 kids for each fertile woman; that means that old people are more and more, while kids and teens won't be so many, so I think I could not find a better way to help others....help new lives to start—the kids are our future!!

I know someone could think there is also a narcissistic element, and I admit there is; I am considered good looking and especially the women of my family (my mum and my sister) are stunning beautiful, so I think the kids will be quite lucky inheriting some of their features. It's nice to think there will be kids with your eyes, your expression, out there and maybe even kids having similarities in psyco. I can't deny there is also this reason, but I almost think sperm donation is something genuine.

If I will ever donate also to a sperm bank, I am sure I will be open to contact; children deserves to know their origins.

As I said before, I cannot imagine a better way to help people—help people to conceive new lives, babies who will be our future.

Hope this message will help to understand better the motivations of donor.

Bye, from Italy

"The Concern of a Donor for the Italian Cryokids"

I am the same Italian sperm donor, but this time I would like to talk not of my story and reasons of being a donor, but about what Italian infertile couples think about gamete donation and the way they will act with their donor-conceived babies.

After starting being a private sperm donor, I decided to join some infertility forum to share my own experience; after a few days, I have just be banned.

The reason was that they don't agree with private donation and for them, donation can only be anonymous in order to not interfere with the parent role.

I read parents writing that they will never tell the children, not because they were scared by their reaction, but because for them, DNA *is of no importance*, genetic roots are nothing as they are the parents and that was all.

Some of them are even just trying to forget they went through egg/sperm donation, as they see it as a *mere medical process.*

Some of them, even if they could choose between anonymous and non-anonymous donor, they just chose the anonymous one to preclude children the chance to know their biological father, just because for them that was just a donor, nothing important....

In my opinion, they are just lying themselves, cause they will never accept they are not biologically related to the babies, so they decide to lie even to the babies and they think this is the best way to "solve the problem," but it's not!

I don't judge badly parents who choose to not tell the child, but what I find horrible is to delete every chance for the baby to know the donor father just in case they discover it!

I admit also the two married women I helped won't tell the children, but if they change their minds or if the children discover it, I am open to meet them....I don't want an offsprings of mine to be a new blogger like the girl of *Confession of a Cryokid*, suffering a lot in missing the half unknown part of herself.

Personally I was *horrified* in reading these forums and the way these parents-to-be act and think, but I read two stories which horrified me the most:

– A woman pregnant thanks to embryo donation write that she will never tell the child and that *if* she decide to do so, she will tell when the son will be like trying and trying to have his own children so to give him eventual medical information. I thought, "What the hell? Use your brain! The son, now a man, is an adult, he is settling down with his wife to have a family, and you traumatize his life revealing such an important truth???"

– An other one went through egg donation and she was saying she will never tell and that she herself was forgetting it cause it was something of no importance, just a stressful medical treatment and nothing more.

Maybe I am too much scientific person, but I don't believe these people saying genetic is of no importance.

I have only a sister, but many, many cousins (nine of first degree, but even seventy only from my mother part), and every time I meet them and all the relatives, we love to talk about similarities in the features, the body, the way we talk and move, cause this give us a stronger sense of identity, and it is beautiful to have such a "big family."

That is why lots of donor conceived just can't wait to meet their half siblings!!

I think in the third millennium, when we are all linked together through Internet, we should learn from the mistakes other did in the past, even if they live on the other part of Atlantic Ocean.

I don't like to think in twenty years there will be the Italian version of blogs like Anonymous Us and *Confession of a Cryokid* with lots of teens suffering and crying, missing half of their genetic roots or even *all* their roots in case of embryo donation.

So I hope this little story can help people in learning from the mistakes of the past.

Lots of love and support to all donor-conceived.

An Italian donor.

"Re: Italian Donor"

[This is a response to "The Concern of a Donor for the Italian Cryokids," found in this volume on page 99.]

Not knowing much about the authors/contributors here, or the author of the referenced post, a legal point is worth noting.

In Italy, artificial insemination with donor sperm has been illegal for a number of years, since, I think, 2008. In fact, thousands of Italian couples go abroad for such treatment each year. This is true. You can confirm my statement easily enough through online sources, including the Italian press.

So how can "Italian Donor" be doing what he states unless it was before 2008? If I recall correctly, a temporary legal ban on the procedure went into effect several years before the referendum. Yes, it was a popular referendum. I remember. I voted in it!

So is he a private donor operating outside a clinical setting? Private donors, acting directly with the recipient, are afforded no immunity from litigation in Italian law, because what they're doing is *illegal*!

Or maybe he is *outside* Italy?

Not to pick on guys in Italy (which includes me), but some of their "boasting" gets tiresome. On the blogs, every Italian guy brags about doing his hot teacher when he was seventeen and she was about thirty. Yeah, right, like *every* female teacher in Italian high schools has sex with her students. Please!

Naturally, none of these claims can be verified.

"Donor"

To all donor-conceived children, I'm sorry. I needed the money, and I thought I was making your parents happy. I thought you would be loved and wanted.

But not by me. Please don't hunt me down and break my real family, for they don't know. My gift of life to you could be my own life's distraction. My real children would not forgive me. My wife, who is my life, would leave me.

I now realise I was wrong. This whole system is wrong. Please forgive me, but I am not your father, nor did I ever intend to be.

"I Am a Sperm Donor"

I am a sperm donor.

And by being that, I am graced to have the chance of having biological descendants, a succession, that otherwise would not be a self-understood part of this human experience called life.

I write this in the knowledge that the likely readers of my lines are affected cohorts of this industry. I am aware of the possibility that you read this eagerly, as a distant message from your own donor, that you would have wished to read.…For when does a sperm donor have a chance to talk, after all??.… The question of the many of you might be, Why? Why have I made this decision of donating my sperm, what were my motives, how do I feel and think of my progeny produced this way? What is the human face, the intention behind the data sheet?

I try to formulate my story from my point of view. Because not only you have a need to talk, to share. I have this need, too. Obviously, I am not presenting anyone else but myself; I cannot speak for any other donors. I guess we all have our own individual stories, but this one is mine.

I am a thirty-year-old single guy, born and living in central Europe. I am 6'1" tall, blonde, blue eyed, considered handsome, some of extra pounds make me a nice cub. Here I go: I am a gay man. And being gay, having accepted myself as such in the past three years, I have absolutely not come to accept being possibly robbed of the biggest gift of life, by design: a family, particularly made up with my own produced children.

It is not the USA here, or other parts of this world of a more open society. I am not saying that it cannot happen, or wouldn't happen, that I will eventually have this dream also come true, but one thing I know for certain: I wanted to make sure that I wouldn't leave this earthly life not having children of my biological makeup. This was basically the "green lamp" amongst a myriad of red lamps that I faced. There are two wishes here, as you see: to have successors, these will be my soon-to-be-conceived children, and to have my own family, with a spouse, and to cocreate my own children, with the help of a surrogate, and raise them; this is yet a very uncertain dream of the future.

One dream I could make come true till this point: since end of 2010, for a half a year, I have donated my sperm several times at a local fertility institute. The samples are being made available during these months ahead, so these progeny are being conceived at this very time and age.

In my country, there is legally only anonymous donorship, no chance of open donating via institutes. I have ambivalent feelings about this. One side of me is relieved, that this is the status quo, every member of the process is aware of it, except "of course" the to-be-born child....I know that, I hear that. On the other hand, never having a chance to see these children of my own....is a hard pill to swallow. I have come a long distance to digest this already, but I guess it has yet its layers, as life will go on. Especially if I eventually will have a family of my own.

As you see, this is the first time I used the word children. I have a hard time figuring out how to relate to them in my mind, in my soul. There are multi-layers of conception here, of course, and I don't even want to diminish the load of it. Would be my self-raised children more my "real" children? In some terms of course they will be. But it would be an insult and ignorance not to acknowledge my otherwise conceived children, in their very existence, having come to existence partly due to my decision to pay that institute a visit....Ignorance never has, and never had good fruits.

The solemn knowledge for me, that there are going to be people on the planet, even whom I won't ever meet, who are my successors. In their eyes you find my eyes, in their smiles, in their very body and even in their psyche, I am there as half of their heritage. This makes me feel that I live in them forward, my very core of being has been passed on, something I also inherited from old ages. I know, there is something deeply narcissistic about this....But this is how I feel. That I have been a vessel of a life giver; it is something deep from my malehood, I guess, too. Mere instinct, made conscious? Maybe.

I have already written more than thirty pages in my diary dedicated to this process, self-reflecting on all the thoughts and emotions that needed to be digested (how gay is that, ha? :)) It was vital to do this....

Is it a solemn, selfish, and irresponsible act of mine, that I freely gave up A, my chance of knowing and raising these children of my own biological makeup, B, their very right to know their biological father, with all the aspects and consequences around this?...Maybe it is. Am I the one to judge? Are you the one to judge?

I genuinely wish for my conceived children the best possible future. Although it is a future I wouldn't be present in and have hardly any power to intervene, to shape....The very constitution of artificial insemination might be a strong proof that their going-to-be parent, parents, really and truly wanted them to be! I think this is a good place to start. Prayer seems to be an ever-powerful tool here, all the power I am left with....I pray to my heavenly Father, that He would grace all my insufficiencies and meet them, and my kids, without the knowledge nor the fathering of their biological father, and fill theirs, right where they are. I wish for you, dear reader, to be filled with grace, too.

What is my ultimate message? I really don't know. I guess we all are put, or have put ourselves, into life situations of insecurities, uncertainty, and places of needs to be met. It is special, in a way, for this group of ours, but also too common in another way....All of us face these trials; literally everyone can relate to this. I personally find strength in my faith in God and am more in need of it than ever. It is good to abandon myself into His good will for my life. He meets my insecurities, insufficiencies, and needs.

What would I say to my conceived children, if they somehow would find me in ten–fifteen–twenty years? Thank you. I love you. I want you to know that you are loved! I am so sorry if I failed you. It was my conscious decision to donate sperm, but it was never my intention to harm you by its consequences. Please forgive me! I love you! Though I cannot see you with my physical eyes, I can see you in my mind's eye and in my heart, and seeing you, I am so proud of you! I. am. very. proud. of. you. Have a wonderful life, darling!

...

This sharing works also as a healing process for me....May it be for you, too!

"Yesterday I Got a Phone Call from the Fertility Institute"

I am the author of "I Am a Sperm Donor." This is my second entry.

Yesterday I got a phone call from the fertility institute, that out of my anonymous sperm donation, two successful pregnancies came to happen recently. And there are still vials left. The call came after my initiated interest in how things are, it was a promised call return. It did not came to me as a shock as it did a half a year ago, when I have had a similar conversation, but it turned out to be false alarm, it was some of an error, and it was not out of my samples the pregnancies had happened. But back then, after hearing that several new lives are on their way who are fathered by my semen, it was more than overwhelming. I was dizzy, I had to sit down really. It all became so surreal. Nothing changed in my life, and suddenly I am confronted with this piece of information, that several pregnancies were successful already. Obviously, as a sperm donor, it is something I was well aware of, that it would be only a matter of time that this would happen; this was the purpose, why I did it, but still, hearing someone tell you on the other line, that babies are on its way out of your semen, and being a single guy, not in this life phase yet to hear something like that normally, it was quite a shock.

Anyway, a couple of weeks later, back then half a year ago, that I called again, it was cleared, that it is not the case. At least not for the time being.

But yesterday it all got real. Two pregnancies are on their way. I failed to ask if there are twins among them, or anything about their conception, I was just holding my cell, and let this information got poured on me. This time I tried, but could not have any emotional response. Nothing. It was in the middle of my work day, actually at lunch break, that I got the call.

Later I googled a picture of a germinating seed, trying to have some focus. I was staring at it, trying to get into the feeling. To have some sort of association. Nothing.

I just told myself, well, later in the evening, laying in my bed, in the dark, the feelings will come. Any sort of an encounter with this information. Actually it was a 1.5-year-long journey since I started writing my diary about this, and contacted the institute for the first time, and I knew, someday had this ending. Someday I would get this call, knowing, that it all got real.

I joined my community group in the evening, we had a nice time together, and then I went home, and fell asleep.

Well, I had a dream. A dream, that got as realistic as only a few of my dreams get. You know those rare occasions in your dreams when you are so into it, and you are getting anxious that this shit is all happening with all its consequences? Such a like.

I was dreaming that two of my friends suddenly have died. Both, who are part of that community group I weekly attend, whom I have met that day. One of them is around forty-five, the other is a college graduate, around twenty-five. The latter looks similar to me, at least I have sympathy for him, he resembles me in some ways, my younger self. I don't remember anymore how they died in my dream, I just got hold of this information somehow, and we were together as a group, and were in shock, but still were trying to talk it through again and again, and to digest the shock, that these young lives will never be here with us anymore. At a point the older guy's persona left the picture, we mourned for the loss of this youngish guy. I was crying in my dream, it felt very intense, very real. It was almost unbearable. I was just circling around this in my dream, that I will never be able to speak with him, and how I will miss him, and how tragic it is to lose such a young life.

I woke in the middle of the night, and slowly realized that it was all a dream. I felt so relieved, that this friend of mine has to be well and alive then.

But lying in the dark, almost crying for real, it slowly dawned on me that this dream was just the emotional reaction I was longing for—it came actually pretty quickly. And I understood what I couldn't figure out during the day, what I feel right now: I feel grief. I am mourning for my lost two children, whom I never will see, never know more about them than the fact that they came to exist. I am mourning my role in the process, that it has just ended where others' is just starting. A new family is about to be born somewhere, due to my cooperation. Joy, and tears of joy are all over I imagine, and here am I, not invited, not to participate in this joy.

I wear black today. God, I am so grateful that I realized, what I am going through from the very beginning: it is going to be a grief process. Now I know it. It is not some winter melancholy, not depression, it is grief. I do hope that this consciousness will help me to get through it in a healthy way.

I am contemplating now about two decisions: I might need to see a therapist to get through this. The other: I might want to open up my donor data on the internet, make a page, or register in some database, in spite of the fact that I am meant to be anonymous. (Here we don't have the option of open donation, only anonymous.) I don't know. It would be still by chance that the recipient mother or later the interested child would find me.

Since these are more U.S.-based databases, I don't even know if the recipient mother, or later the child, if they want to find me as the donor, they would even think of me being registered in a U.S. database, being a central European.

Anyway, I am aware that all this is part of the mourning process. I guess this is some glimpse of hope, that I could be reached, get involved if they want it like that. I am open for this. Still I am not sure if it's going to hurt them, or hurt me—I guess I have time to make up my mind around this.

I don't know anymore, what I imagined it to be like, when I first decided to donate my sperm, how the process would evolve, but now actually it is somewhat painful.

Anonymous Sperm Donor

Afterword: I guess the reason I share this here is partly due to the common notion about sperm donors I see in writings, that he surely is some irresponsible college student, who will buy beer for his next party out of the money he gets for jerking off into a cup, and afterwards he would never think of it at all. A selfish pleasure seeker. That he simply doesn't care. Well, this is absolutely not who I am. It kind of hurts to be perceived like that, or even anything near that. I don't want an award; don't get me wrong—I just wanted to bring in another aspect. I guess I wrote more about my feelings in my previous entry.

"The Many Different Feelings of Different Donors"

[This is a response to "I Am a Sperm Donor," found in this volume on p. 102.]

Sperm Donor, I read your story.

As a sperm donor and someone knowing many sperm donors via Internet, I can say such these feelings are more frequent than anyone could think.

There is a donor in my country that donated to a sperm bank like you and then privately too: now he is o.k. and happy about it, but at the beginning he was feeling even guilty. He wrote to a forum: "Am I doing the right thing? Sometimes I feel like the cuckoo bird who leave his eggs to the other birds letting them raise the offsprings...."

Another one who knows he fathered twenty-eight felt overused, and even me sometimes, I wonder some questions like, "Wait, but these will be my biological kids...and they are far away from here....It sounds weird," but then in my case I am so happy for the parents and for the children created that I am almost enthusiastic about donating.

Everyone is different though and have different feelings; what I suggest you is to ask the clinic if they can write somewhere you could be open to contact....Maybe some kids will call the clinic in the future and if you say you are open maybe they could call you....or yes you could sign up to a special site like DSR.

Bye, and support from a sperm donor

P.S. If you want to talk about your feelings with other donors, sign up on Freespermdonorregistry, which will be soon KnownDonorRegistry....I am sure you will feel better!

"Happy Egg Donor"

I donated eggs right before starting professional school years ago. I'm happy that I was able to help a couple have the children they wanted so very much. I'm curious about the children who were born through this process but wouldn't try to intrude into their lives. The parents who have children using donor sperm/eggs don't want their maternal/paternal claim to the resulting children undermined by the donor. Egg/sperm donation, when done right, is very much a positive thing. It isn't right that infertile men and women should be prevented from having children just because of a biological issue that they have no control over.

If the children wanted to track me down one day after they'd grown up I would give them the level of involvement that they and their families were comfortable with. I feel like I should do right by them/their families because they were born of my eggs and because of the arrangement that was made with the parents.

"Proud Donor Wanting to Offer Info to Children"

I was given up for adoption at birth because my mother was sixteen and my father had been killed in Vietnam. But I have since met my birth mother and learnt my adoption was not due to a lack of love or desire to be a parent, but it was a lack of age, money, and family support, *but* she had an abundance of love for her child that she wanted me to have more than she thought she could give.…I am so proud of her for being strong enough to protect me whilst hurting herself.

I grew up in an open, loving adopted family and was given everything I needed and wanted. I was told early on that I was adopted, and as a child, I spent years fantasizing about having rich, famous parents or rock star siblings.… then as I got older, I started being more curious about whose smile, humour, good/bad looks I inherited and what my biological parents were like. Because my biological father (an only child) and his parents had died, I then started to research his family history and genealogy to learn who they were.

Somewhere along the way I felt the need to have my own children. I had a very happy life, a great experience with finding my birth mother and her family and then finally knowing who I was.…Only problem was that I was thirty-six and not married or in a relationship.

Sure my story is all sweet and rosy, but why shouldn't another child be given the opportunity to grow up in nonconventional family like I did and be allowed the possibility of all the love and joy I received.

So I became a sperm donor. *But* I wanted my child/children to know as much about me as I could, so I practically submitted a full dossier of my history, personality, looks, family history, pictures, and a letter to be given to the child/children introducing myself, telling them why I was doing this and offering the mother/parents that I was prepared to be as involved as much as they would let me in the child's life, from birth onwards and without question or claim. I am not sure if my "dossier" was ever passed onto any parents/children, as its only been a few years.…but I am one donor that would welcome the opportunity to help any child born through my donation to gain that ultimate prize, the knowledge of who they are and where they're from. Only adopted/donor children will truly understand the wealth and value of knowing who you are and where you come from.…

"Concern for Potential Parents-to-be"

When I was younger, I never thought I would be an egg donor, let alone surrogate. I went through the egg-donor process four different times to help two families have the babies of their dreams. After losing my own child, I became a surrogate, and after another loss, finally am waiting to give birth to my first surrobaby.

I cannot explain what compelled me to do this, but here I am. I am not necessarily writing about my own story but about a concern I have.

I *know* of a woman (I can give the name/email if I am allowed to) who wants to be an egg donor for *all* the wrong reasons. She is not a good person, she has told stories of issues she's had, and I am very concerned for potential IPs [intended parents]. Unfortunately, I don't know where to go in order to ensure that someone doesn't inadvertently get screwed by this lady. She acts like she's a nice person, but is very much the opposite. She will do anything for a buck, and that includes to *lie*. She is *getting a divorce* but is trying to donate her eggs and get money so she can move. She's not being entirely honest about her history and potential problems that can arise.

I don't know how to potentially stop her. She wants to be an independent egg donor (she would not get approved by an agency). Somehow, I feel strongly that this will end badly for all who get involved with her.

"So Wonderful!"

Even in high school, I knew I always wanted to have children and felt sympathy for the couples who tried to conceive but couldn't. Finally when I finished college, I felt it was time to give something that I knew would be the best gift I could ever give someone, and that was the gift of life. My decision was based primarily on the gift I gave and not on the financial compensation. Sounds altruistic, but I had thought about this decision for years.

As a redhead with a good education, I was snatched up right away by an anonymous couple at an IVF clinic. I went through the treatments and became overstimulated with twenty-four eggs! But with all those eggs, the couple would have the chance to have plenty of kids! I got a call the following year, saying that I "did a good job!" so I decided to do it again. And then yet again, the following year. Three full cycles for me for three consecutive years.

Donating my eggs was one of the most satisfying things I have done in my life. Knowing that I gave life to someone who otherwise was unable, brings me great happiness. The only tangible memory from these three donations is the beautiful silver heart pin the second recipient gave me. Filled in on one side but open on the other, the center has a small gemstone representing the egg. I cherish this gift and wear it close to my heart. I wish she knew how much that gift means to me.

The children who have my genes are not my children. They are my genes. Their mother is their mother, as is their father (a sentiment that my husband doesn't share.) I now have two children of my own and know firsthand now, unlike when I was in school or as a donor, what it truly feels like to be a mother. Life would be completely empty and void of meaning without children.

I am proud of my choices and would be interested in meeting these children at some point, given that they and their parents would like to meet me.

"It All Started Out with Lies"

I was in my midthirties, realizing that I might never have children of my own. It all started with me just browsing the search engines and came across some sperm-donor forums ads. I remember reading one of the ads—I thought, this guy is full of it and lying about everything. Then I realized a majority of the donors in some of these online forums were lying about their education, height, how many they fathered, type of employment, etc. There were so many donors who just wanted to have sex. Maybe cause they had problems getting sex from females in their normal life.

Many of the recipients were so desperate in finding a donor they sometimes had sex with these donors. Even if these guys were infertile and never fathered any children. They usually make up a number of children they father and the females would often believe it. I used to sleep around with females from the bars and had a very reckless lifestyle. I started out lying just like some of these guys. Guess the only thing I was honest about was the way I look and that I'm STD free. I'm more honest to the females nowadays. As for why I'm a sperm donor—basically I couldn't afford to have children of my own. I work a low-paying job like so many people. I did enjoy being told that a recipient is pregnant and how happy she is.

"Relief from the Real Problems, Followed by Regret"

When I was in high school, I had a friend who had a meth problem. She carried bags of gummy candy everywhere she went and drank a lot of Diet Coke. She read fashion magazines and got brilliant grades. I would get sms messages from her sent at 4:00 in the morning on a regular basis. She got scarily skinny in a short amount of time. People started to talk. I started to worry about her. I could smell the meth on her. I told her she needed to stop. I threatened I would stop being her friend.

One day she broke down to me and told me that the meth was her relief for her real problem—bulimia. When she smoked it, she felt like she didn't have to purge because she knew the drugs were burning off the calories. I had no idea she was bulimic. The only evil I saw was the meth—but it was just her resolution to a problem—feeling not good enough.

It was very important to my mom that I go to university, and I didn't want to disappoint her. But she didn't have any money, and my dad was not in the picture. In the rural area I grew up in, there were not many opportunities. So I paid my own way through university—only I used student loans and hoped to pay them later.

I worked retail and waited tables and I still couldn't keep up with my peers. Rent, clothing, entertainment, travel expenses, cell phone, car costs, and student loans just kept adding up. I went on Craigslist and started looking for extra work. Most of the jobs didn't suit me at all. I either had no interest in them or no ability to pull them off. Because I was unskilled, my pool of job choices was very small.

There were several men offering money for "gentlemen's arrangements"—cash for a young pretty female to spend time with them. It was a euphemism for prostitution.

When I "donated" my eggs, I did it because I needed the money. Also, it was a relief for me. To support myself, I had been sleeping with a married man with five kids for several months on a weekly basis in exchange for this "arrangement." Finally I couldn't take it anymore. I still needed the money though, that's when I sold my eggs. On the application, they ask, "Have you ever exchanged money for sex or drugs?" I lied and said, "No."

I had a decision. I could sell my reproductive parts either to old men or old women. The latter was supposed to be a relief to the former. I have a young child out there now that I can't see or know because of my selfishness and my inability to feel "good enough" without a degree, without the clothes, without the car....

I'm sorry to the child and I'm sorry to the wife of the man I slept with. I truly am sorry.

"Joys of Being a Surrogate"

I wanted to stay home with my two young boys, but I needed to find a way to pay the bills. I saw a story on Global National about a surrogate; she was a homemaker and wanted to give a couple the gift of a child. I have been thinking about surrogacy for some time, but didn't know much about it. I started to research online about the "laws" surrounding becoming a surrogate, and tried to find a couple looking for a surrogate online.

Everything fell into place, the head of two fertility consultants contacted me via email. One really stood out in my mind; she is not only the head of the company but she also, like me and others, was a surrogate and knew what I was looking for. We talked on the phone a couple of times, and in a matter of a week, she had found me an IP (intended parents) couple to work with.

We had an appointment at a clinic downtown, and I was so nervous that they wouldn't like me. Only one of the IPs could meet me that day, but the fertility consultant also met with me to make things easier. He (IP) walked in and I hugged him right off the bat, you can tell he was like, "O.k., I don't even know you." LOL. But he was so nice, and he smiled a lot, like me. He was funny and joked around like I like to do. He asked about my boys, and we had a good first meeting.

We had appointment after appointment. The transfer date came, and I was so scared. It was a couple of days before Christmas, and all I could think about was how sad and upset I would be if things didn't work out. The two-week wait, better known as the 2ww, was long and scary. We didn't take a HPT (home pregnancy test) so the IPs didn't get a false positive or negative. I prayed every day for this, and I didn't want to start the Christmas holidays off with bad news.

2ww was finally over, and the other IP and I got the great news that they were pregnant. We hugged like a million times, I had to hold back my tears. I was so happy for them both. We are now thirteen, almost fourteen weeks pregnant with twins, and I pray to *God* every day, I'm so thankful to have met such great IPs and that they are blessed with twins.

I think if you are a young person that is done having kids of your own, and you want to help someone out, this is one of the most beautiful gifts you can give. All b.s. aside, I wish and hope that I would have met the IPs in school, or at a party. They are the most loving people I know, they're so in love. Our relationships are so alike it's not funny.

I hope nothing but well wishes to them both, and their beautiful twins. I know after the twins are born the IPs will be sooooooo busy. I know that we may never talk again and I have come to terms with that. It's hard in some ways, cause I'm a very emotional person, and I care for the IPs, but I knew when I started this that there was a chance that we may never talk again. I'm just thankful that I got to meet such wonderful people like the IPs no matter how short of a time. They will always be close to my heart.

Surrogate

"Joyful Egg Donor"

I donated my eggs while in graduate school, initially in response to a radio advertisement. I liked the idea of helping a couple build their family, since this is something my husband and I wanted but had not yet started, and the money would help support us.

The agency stressed the importance of the process being anonymous, though at this time I can't remember the reasons. I made it through the screenings, and fairly soon after, a couple chose my profile. I believe it was during the screening process that I indicated that I would like to give the couple my genealogical history, since I have done tremendous research on the topic, and the couple accepted that offer. As we progressed through the process, I prepared a CD with pdf files of my ancestry, starting with my great-grandparents in order to keep it fairly anonymous. It was always my thought, though, that with all the genealogical information on the web, that the child could use that information to at least greatly narrow down the list of possible donors.

I had given the couple an email address and received word when their daughter was born, along with a card through the donor agency. And since they were able to collect about a dozen eggs, I was told that they could have additional children as well. I have not heard anything since that initial announcement but hope all is well and that they have a nice family of their own.

I have since had two children and have completed my family. I think about my other biological children occasionally with happy thoughts of how I was able to contribute to their life. If I were contacted, I would be happy to meet with the family and would open my life to them if they wanted or needed.

"Contacted Donor"

About twenty years ago, my wife and I were trying to start a family, and after several months of unsuccessful attempts, I had my sperm checked at a hospital, which did IVF treatment. The hospital advised that I was okay and asked if I would become a sperm donor. I agreed to the request, as it would help people have a family and I had some experience with the emotions that come into play when the desire to have a family is unfulfilled.

Fast forward 19.5 years to Sept 2010. A letter arrived informing me that I had a child conceived from sperm I had donated, and she wished to meet her biological parent and would I agree to the request. My immediate response was yes, and after discussing the matter with my wife, we met with a counsellor at the IVF clinic and then arranged to meet the child.

Meeting the child (Sophia) was a very positive experience for us and for her. We organised the meeting through her mother, as Sophia was only fifteen at the time. We arranged a second meeting with Sophia and her mother at our house to enable our children, a boy aged twenty and a girl aged seventeen, to meet Sophia and her mother. This meeting also went well, and we have had subsequent meetings with Sophia and her mother, which have been positive and enjoyable.

To illustrate how well the contact has been received, my daughter has requested that I ask the IVF clinic to write to the families of the other children (seventeen) conceived by my donations to inform them that I am okay with being contacted if that is what the children wish to do.

My attitude to meeting with the children is one of intellectual curiosity. I am not their parent, I am someone who provided seed material to their parents to enable them to grow their family and raise their children.

My experience and message to donors is that a child who wishes to make contact just wants to put a face and some facts to their history and get on with their life.

"Finding the Right Balance"

I'm a donor, both private and clinic. Where I come from, the donor is often the legal parent, so anonymity is advised. This is not ideal!

I don't think there's anything "wrong" with sperm donation, but I think that honesty is the best policy for helping donor children cope with their origins. It seems wise for mums to be open from the start, to help avoid any feelings of deception when children discover their origins.

Personally, I'd argue for the maximum availability of donors for later contact, should this be needed. Also, I accept the fact that some donors feel they can't donate without being fully uncontactable, as opposed to just anonymous, so personal choice in the matter should be available in order to ensure that the supply of donors is not adversely affected.

Frankly, I believe the shortage of donor sperm would be a lot less of a problem if there were better protection for donors from vexations and mercenary litigation. I feel the best way is for donors to be open and known to the families. This can only come about if with proper legal protection for donors from child maintenance claims, and also with legal protection for their privacy.

If we could move to this system, as I understand is the case in certain parts of Europe, I believe that the whole system would work better for everyone. With better protection and the option of some contact, more men would be persuaded to assist—and for better reasons than a few dollars from a sperm bank.

Being a donor is also very hard emotionally, and I also think that these changes would allow donors to cope much better with the doubts, confusion, and emotional fallout of having a number of children that are not part of their family.

Donor conception isn't going away, so let's be grownup about it. An open system of contact, with full privacy and financial protection for all recipients, would be the best system.

"Lies and Betrayal"

I'm a father, a "known" sperm donor. I was asked by a lesbian couple, people I considered dear and trustworthy friends, to help them conceive two children. After discussing it at length with my wife, I agreed to help them. In short, our agreement was that I would consent to adoption by the nonbirth mother. In return, I would have as much involvement in the children's lives as I desired.

I fell in love with my daughter the moment I saw her (in truth I loved her long before she was born). That's when the first stirrings of trouble began.

Turns out my "friends" didn't like my attachment to my child and started making excuses for why I couldn't see her, almost immediately. The stress of the situation gradually escalated and started negatively affecting my marriage as well.

Despite clues to the contrary, I continued over the course of the next 2.5 years to help my "friends" conceive a second child (fortunately that never happened). During this time I became more and more attached to my daughter, and she to me. Finally, my "friends" gave up the illusion of civility and severed our relationship, along with my relationship with my daughter. My marriage also ended around this time.

Fast forward to the present. I haven't seen my child in several years. Any attempt I make to contact her is met with verbal abuse and legal threats from my "friends." My daughter has no idea she has a father, let alone who I am. Ignoring persistent legal threats, I've made myself easily found on the Internet. When and if my daughter tries to find me, it will take her fifteen seconds.

It has taken me several years to begin to accept the fact that I was used, lied to, and betrayed. I was a means to their end, nothing more. The twist in the ongoing story is that I'm not going anywhere. I'm easily found, and my arms are wide open to my daughter. If she wants a relationship with me, she will have it. I pray that she does.

My advice to anyone considering sperm donation is this:

1. Don't.

2. If you insist on doing it, don't take it lightly. Get an attorney, and get everything clearly in writing. If you ever want to see your child, get those details in the contract. Don't leave anything to chance. If you think it's not necessary between "friends," you're wrong. People lie sometimes.

Maybe not all such arrangements turn out so horribly as this. But in several arrangements I'm aware of, there is some amount of lying going on to family, friends, and child. Your mileage may vary.

Regardless of what anyone says, you are the parent of that child forever. Your decision has life-long consequences, and no amount of rationalizing will make that go away.

If you are a donor-conceived child, I'm sure that your birth father and/or mother are eager to see you. You have a right to know them, and you have right to know you're loved.

This entire "friendship," this entire arrangement was doomed from the beginning, because it was based on lies. I wouldn't change the fact of my daughter's existence for anything, but the actions of my "friends" have caused a great deal of harm. In my case I made a promise, and I followed through on that promise. Some people would call that stupid, and I can't totally disagree. But I still have my integrity. I acted in good faith.

To my daughter: I want you to know that I love you and that I long to see you.

You are on my mind 24/7. You are in my life every day, and you are part of me. Nothing will ever change that.

I love you,

Your father

"An Unselfish Act of Love"

When I was a senior at Harvard working on my perfect 4.0 GPA, I responded to an ad seeking egg donors. Since I was on scholarship and Harvard is extremely expensive, it seemed like the answer to my prayer. I made an appointment and submitted myself to the battery of psychological and physical tests to see if I could qualify. Lo and behold, I fit the specific criteria they were seeking. While most egg donors are blond hair, blue eye, I am the product of biracial ethnicity, and it so happened that is what the client was seeking.

No need to go into details, as most everyone involved in donation knows how the process works. Since I am blessed with a high IQ and I am also multilingual, I was able to command top dollar at the time: $8,765. I always wondered how they came up with that amount. In any case, all went as according to plan and soon after, here comes baby. The money was a blessing, and I was able to support myself for a while. When I graduated, the fabulous career I envisioned did not materialize as fast as I had hoped. Call it luck or serendipity, but I received a letter from the agency wanting to know if I would consider donating to the same couple who now wanted a sibling for their first baby. Since the real world now beckoned with bills, car notes, and rent, I eagerly agreed, but I negotiated a larger price this time. I asked for $23,000, and the family, who by now was extremely wealthy, quickly paid it.

The money allowed me to move back to my hometown in the South and get on with my life. I can honestly say I do not feel I have any "children" out there. Nor do I lose any sleep or suffer angst-filled nights lamenting on what might have been. I was simply a conduit that allowed a loving couple to achieve their goal of having a family. I have never wanted to know the girls or, by now, young ladies, as they would seem like strangers. Their lives belong to their real family, and if somehow I get a card from the agency asking for anything more than medical information, I would have to respectfully decline.

If I were still able to donate, I would absolutely do it again. The agency showed me pictures of the parents as they met their children for the first time. How anyone can be critical of that is beyond reason.

"Oops, I Sold My Kid Away..."

When I was twenty I was in college and looking for a job that would be easy—I had a lot of obligations, not a lot of time, and not a lot of life experience. I wanted everything, and I wanted it now, but restaurant and retail jobs didn't pay the kind of income to buy me everything I thought I needed.

A funny thing happens when a girl becomes a woman—the world starts treating her as an adult; middle-aged men begin to ask for dates, people stop forgiving you for your stupidity and start expecting you to behave responsibly, and suddenly you're allowed to do whatever you want. "Whatever you want" sounds like a lot of fun. Well..."whatever you want" sometimes isn't "what you need," and mistakes become profound, even monumental.

I popped open my laptop and began my search for the perfect easy job. For my lack of degree and vocational skills the pickins were slim, but for a pretty young girl willing to compromise her integrity, there were plenty of ads promising fast cash. I could be a candy-stripe girl, a discount masseuse, a foot-fetish model, a dog walker, babysitter, marketing-research participant, pharmaceuticals test subject, or nude model.

"Thanks, but no thanks," I said, until I came across that seductive "$10,000 & Up!" ad. "Give the Gift of Life!" they said. No kidding, with *that* kind of money, I could actually have a life....

I signed up and began what would be one of the most dehumanizing experiences of my life. They wanted pictures of me, lots of pictures, and medical information, and personal information. They only wanted baby pictures. They weren't interested in the adult their child might become; they wanted the baby version of me. It was obvious there were right answers and wrong answers in the questionnaire. The more generic you could answer a question, the better. Intended parents want to project themselves upon your DNA. Any kind of real personality is a turnoff. Bad girls need not apply. And then came the hormones….Ouch.

Somehow I made the cut, and there is a kid out there somewhere I have no ability to contact. It didn't hit me until I called them and asked if there was a birth. "Yes" is all they told me. Sometimes strangers and acquaintances ask me if I have kids. I want to tell them "yes." But "yes" is all I'd be able to say. I wouldn't be able to answer if it were a boy or a girl, or how old, or what's their name, or when their birthday is—just a vague and bewildering "yes."

I don't even remember what I spent the money on. Debt, dresses, and nice dinners probably. I'd give you $10,000 this very second for the chance to meet my kid. Biggest oops of my life.

14. Personal Stories from Adoptees

"What My Mother Told Me"

My natural mother was a beautiful "oriental looking" woman (as the social worker described her), with a natural flip of hair, who left me at the hospital. She told me that a friend of hers had asked to take me, but she said "no." Another friend, a doctor (not my father), asked her to marry him. "No. I didn't have those feelings for him." The doctors who treated her while she was pregnant told her to "keep the baby." She always said "no." She said that she didn't want to be poor.

When I met her, she also said, "I wish you found me earlier, when I might have been able to do something."

She left me, but hearing her voice is relaxing to me. I love talking with her… even being near her.

"Desperate for a Child"

When I was very small, my mother told me my dad was not my biological dad. He did adopt me when I was two and has been the only dad I know. I don't remember how she told me, but it was always in the back of my mind. When I was twelve or thirteen, I wanted to know who my biological father was. I referred to him as "the sperm donor." I really don't recall much of what my mother told me, but I began to be a very difficult child.

When I was little, I was very close to my dad, but when I learned that he was not my biological father, I was very hurt. I learned later from my mother that I would always tell her that I didn't think my dad loved me. He did love me, but I think he was also hurting for me. My mother told him that I was not his biological daughter and he had no right to discipline me. This was a huge problem especially when I got older, because I longed for that relationship with my dad. I resented my mother for this.

My mother was on her second marriage and already had two kids from her first marriage. She wanted another child, and her second husband could not have children. He told her to go ahead and go find someone to make a baby with. How could a husband tell his wife to go out and do such a thing? To this day I can't understand why.

My mother found a man about ten years younger than her that she worked with to hang out with. He was also married with children. He was in it for the fun, and my mother was in it to have another child. She says she told him this from the very beginning. His only concern was would she ask for child support. She promised him no. Through this arrangement, I was conceived. His wife didn't know anything about the relationship he had with my mother or that she had gotten pregnant.

I was born and brought in under my mother's second husband's last name. I think my mother said that the sperm donor saw me once when I was born. After a short time, my mother divorced her second husband and met my dad. They did marry and as I said, he adopted me. They are still together today and very much in love. I am glad for this. When I became an adult, my mother told me that dad always told her if they ever got a divorce he would fight for custody. I think dad felt mother was wrong for not allowing him to partake in my upbringing as a parent should. By rights he was and is my only dad.

For a long time I felt dirty and unloved. I couldn't understand how my mother could bring a child into this world like that. I resented her decision and have treated her poorly.

I now realize that my mother wanted me and planned for me, just not in the best way. We are working on our relationship, and I have since become a Christian, and through the Lord I am healing.

I did meet the sperm donor when I was eighteen years old, and it was a big disappointment. He was a drunk and lived a crazy life. Shortly after meeting him, he up and moved and didn't tell me. Then a few years later, I sent a Christmas card to the sperm donor's mother, and she returned my card with his obituary. It was quite devastating.

I am now thirty-three and beginning to learn to love my mother and forgive those who have hurt me. We have a long way to go, but I can accept that I am loved and was not a mistake.

"They Said It Was the Best Placement They'd Ever Had"

Because I was a mixed-race child in the 1950s, I was hard to place. The social workers looked hard and found two families that they felt would qualify. The family that won out had already adopted a child who was the same sort of mixed race…Hispanic and Japanese. It is a rare mix that I have yet to duplicate.

My parents-to-be came to the meeting with me and the social workers, along with my brother-to-be. He brought along a football to play catch. He was twelve, and I was a little over one. The social workers were so impressed at him. At the same meeting, I fell asleep in my mother's lap and everyone was amazed how calm I was.

The next meeting, my parents took me home, and the social workers said how wonderful a placement this was… the best they had ever seen.

"Isolation"

Being involved in a church community, school…having friends…trying to communicate with my parents…I was essentially alone regarding being adopted. I probably knew some other adoptees. But adoption was never a topic of discussion. When it did come up, there was nothing much to say. The silences lasted.

Sexuality was tough. I kept thinking about my father who left me and my mother…and my mother who left me. Every time I was sexually attracted to someone, I thought that this was dangerous. I never knew what to do about those thoughts and feelings. I was self-conscious and anxious.

I was anxious a lot, so I self-soothed when I went to bed. I don't know where I learned it, but I did it a lot even when I was very young. I don't think it helped me develop sexually. My fantasy life was stimulated by it, but my relationships weren't helped. I never learned to talk about it. It was a secret. I was good at secrets.

My body was strange. My skin was darker than almost everyone. I was a different race, and to me my face was shaped like a monkey face. All of these odd facts made me feel isolated and different. I imagined myself living alone in a trailer as an adult.

But these thoughts ran counter to others. I also wanted to live and do things. I loved beautiful things, for example, watching far-off thunderstorms, the colors in the evening sky. I wanted to be as expressive as the sky, and all my secrets, hidden darkness, to be gently treated, held, loved—and told that they were okay and normal and good: my body, my skin, my sexuality accepted.

It was a discovery that these shaped my life. Trying to conform and fit like others, I was shaped like clay by adoption's fingers—by secrets. Silences pressed, creased, accented. So much in private, hidden from me. I know things about solitude. I'm still trying to find words.

"It Was About Truth"

When I was forty-eight years old, I went to a friend's wedding. Afterwards the priest came up and asked me if I was Hawaiian. I smiled and said, "Yes." Later, driving home with my wife, I said, "I don't know if I'm really that or not."

I had gotten to the point that I wondered if my parents really knew where I came from. I was adopted when I was fourteen months old, after being in foster care. I was a mixed-race child, so I was hard to place. Social workers took care of my files. How do I know that the social workers told them the truth?

I remembered that my dad said something just before he died. "You might want to look up your natural mother, she came from a leading family in Hawaii." At the time it just raised more questions in me. He is dying and he is telling me something about my mother. Was he just saying something to make me feel better? Was this just one of his fantasies? Or was he saying that he suspected I might want to know her?

It's funny that I thought my dad might say something just to make me feel better, as if I needed comforting or would believe whatever he said. He wasn't the sort of person to lie. But I always thought he was a little sentimental. Maybe he would shade the truth in brighter colors. After he and my mom died, I wondered what was true about where I came from.

After spending lots of money—probably close to one thousand dollars—I got close enough to my mother to have to hold my breath for two weeks. I had to wait for her permission to arrive at the court in the county of my adoption before I could learn her name.

Yes, I was conceived in Hawaii. Yes, my mother and father were (pretty much) who I was told. My parents had that right, and I was glad, because I wanted to believe them. It bothered me that I doubted. The truth is something I had to know for myself. In fact, the truth was much more interesting than anything I imagined. It was the dreams and fantasies that got to me. What right does an adopted person have to the truth? The same as everyone.

"Don't Drive Yourself Crazy"

As an adoptee, it makes me sad to read about the grief many adoptees and sperm-donor children are feeling. The saddest part about it is that I think a lot of it is self-induced.

I know that growing up, there were times that I felt like my parents didn't understand me and that my "real" parents would have. But, the truth is, every kid feels misunderstood growing up. The difference is, adoptees/donees have a fantasy world that they can retreat to where everything would be easier if they were with their "real" parents.

If children being raised by the people that gave birth to them often feel misunderstood, why in the world would adoptees/donees have it any better if they were with their birth parents?

I also have the perspective now of raising my own children. I'm sure that they have felt misunderstood on occasion, and the truth is, there are times that I have not been able to understand them. We are creatures of our own inherent personality and our environment. Our personality or temperament may be similar to that of our "real" parents, but then again, it may be completely opposite. I know that most of you have heard this before, but it is true that the people who have raised you are your real parents.

Find your strength from inside. Adoptees/donees who were raised by loving parents only have a "hole" or a "missing piece" if they decide to create that hole themselves.

"The Missing"

I was given much, and gained much. Outside in, it was a dream world; seldom have two people wanted a child so much, or been so prepared to love one. Daddy spun me round, taught me science. Mom nurtured, taught me nature. A dog, a cat, a horse. Room to run. Private schools. Swim team, Pony Club. Charmed life, I'm sure. Who wants to hear that the princess wants out—not just out of the tower, but out of the fairy tale itself? I didn't voice the pain. I didn't know who I was, or where I came from. What an insignificant burden to bear in a world filled with pain.

Curiosity mixed with longing becomes an itch, and the more I scratched, the more I searched, the more I uncovered a wound. Shut out, locked out, told I had no right to know, I railed against the walls that had been erected around my identity. It never leaves you. People unknowing, unthinkingly, pour on salt, pick, pink the scars. A request for a medical history. A reciting of the family genealogy. A story—a birth story—recited to a son on his birthday, knowing that my own was lost. It's minor, I told myself. It's a flesh wound. But need outweighed gratitude, and I searched again, and found.

Prepared for the worst, for rejection, for shame, for recoil, I was given welcome. They spoke of longing, never quelled. Of whispered prayers and silent meditation. They missed me, all those years. That was the gain. A family history. A flick of a wrist that resembled mine. A way of seeing the world that was shared.

I wish it didn't hurt. I am grateful. But the picture of my other parents, her arms empty, can bring me to my knees. I am missing. An obituary, the count just slightly off, less than truth. I am missing. My great-grandmother's ninetieth birthday, with the whole family gathered around.…I am missing. Always missing, for thirty-some years.

My eighteenth birthday came and went, and half a continent away, a grandfather I should have known breathed his last. The stories that are hardest to hear are the ones where I'm missing. What came before, in the decades and centuries before the 1970s, is my birthright. They don't bring pain, and they aren't hard to hear. But the ones between then and now cut like a knife, because they're part of a life I didn't lead, where I was silent as a ghost, and no more substantial than a wisp of air—because I was missing.

"Finding Me in the Bathroom"

Today at church I got this "revelation" about adoptees. We're not really able to publicly grieve or even acknowledge the pain hidden deep inside over the loss of our mothers or our original selves. No, we *have a mother*—we're "special" and "chosen"—*why would we need to grieve anything*? Weirdos.

So I got up from my seat after church (wiping away tears—I'm a big fat crybaby), and hightailed it quickly into the restroom to "freshen" up. I hurried past the rest of church society intermingling, conversing, sharing, connecting—and made a beeline for the bathroom, as usual. Didn't want them to see me completely undone emotionally.

You know why? Because then I'd have to be *real*. And what I realized is that most of society *doesn't want* the *real me*. They want the "Happy to be Adopted" me. The one who, sure, may have a *hint* of *curiosity*, but nothing that really affects me deeply—because I am fulfilling the role that was written out on my amended birth certificate, my adoption decree—that is *who I am supposed to be*.

Society buys into this and shrinks back from any adoptee who breaks free (even through the deep help of the Holy Spirit) and dares to become *real*—not just the person they were legally created to be—but who God created them to be. And embrace. And heal. And live. But many times, it has to be done alone. And in secret, like sealed records in the courthouse. Adoption is like that somehow.

It creates *lies* to *cover* truth— true parentage, true identities, true stories, true realities. They are replaced by man-made truths— new parentage, new identities, new stories, new "realities." New certificates. New names. New homes. New lives.

But nothing man-made lasts. It eventually crumbles, and *truth* prevails. Even through great pain. Kind of like the cross.

I'm so glad that someone wants to know the real me. Even if I am in a bathroom stall crying, seemingly alone.

15. Personal Stories from Fertility Industry Professionals

"Health of IVF Adults"

Recently, as a professional ethicist, I attended a lecture on perinconception health in which evidence was presented on health risks generated by what happens to the parents from twelve weeks before conception until eight weeks of pregnancy. One of the issues was the effect of obesity in either parent reducing genetic effects that predispose to obesity in the child, even if the family did not have a history of obesity beforehand. Some early results were discussed of a major project to assess the health of IVF adults. It was said that there was strong evidence of insulin resistance in IVF adults. That predisposes to a range of adult disorders including diabetes and polycystic ovarian disease in women. It was suggested that the insulin resistance might have something to do with the culture media used for embryos. I am aware of a project involving 93 percent of IVF adults conceived in the Australian State of Victoria. I would be interested in any other long-term follow-up projects assessing the adult health of IVF-conceived people.

"We Have to Change"

I am a mental-health professional specializing in working with infertility patients. I am also an adoptive mother. Part of what I do at work is prepare people for the implications of using a donor. The most challenging topic we discuss is the issue of whether the couple will tell their child how s/he was conceived.

As an adoptive mother, I am 100 percent pro-disclosure. I feel it is a child's fundamental right to know who they are at every level. Yes, it's confusing as a parent to know how much or when to disclose, but that confusion can be eradicated with some solid education and preparation. Part of what you sign up for when you build your family in a nontraditional way is that you will have to parent your child about how to deal with that difference.

To be honest, I myself could not imagine using a donor, because I don't understand the intense need for genetic link or to be pregnant. Yet, I find myself in the position of counseling many individuals who have chosen a very different path from the one I have chosen in life. At times I have even considered shifting the focus of my practice away from infertility so that I can feel more comfortable with my work. Then I think of the importance of the discussions I have with people and rationalize that at least I am a voice for donor-conceived children. At least, I bring their voice into the picture before couples even become pregnant.

I am amazed at the number of clinics, physicians, and agencies that really do not give even one second of thought to the needs of the children that are being conceived via "third party reproduction." As a nation, we need to get a conscience about what we are doing here. Yes, it's nice when an infertile couple is able to build a family, but what about the children? Shouldn't their needs be in the mix from the very beginning too? I think it is ridiculous that a donor-conceived child would need to "research" to find out their genetic origins. Give me a break. What if you had to do that? Is it fair?

I would like to say that most of the mental health professionals who specialize in infertility really do understand the complexity of these issues and support openness around donor conception. I feel proud of that and feel that we are having an impact on couples that come through our offices. However, until all couples and donors are required to consult with us so we have a chance to open everyone's eyes early in the process, it will be an uphill battle. I conduct my consultations with donors and couples using ASRM's ethical guidelines for third-party reproduction. The guidelines are solid, but not mandatory, so if a clinic or agency chooses not to adhere to them, there is nothing anyone can do. Consequently, many people go into arrangements with donors pretty clueless about parenting issues and implications for their donor-conceived offspring. We need to change this situation. It's not fair to donors, intended parents, and especially to donor-conceived children.

16. Personal Stories from Parents

"Access Denied"

Fourteen years ago, I was in a new marriage, with a wonderful husband and two gorgeous children from my previous marriage, and I was on the threshold of acting upon the yearnings to have a child with my new husband, especially as he was such an amazing stepfather. I felt he deserved the right to have a child of his own.

Firstly, as I had had no problem at all becoming pregnant when I was younger and in my first marriage, I didn't foresee that there would be any issues this time.

I did fall pregnant a couple of times. However, these pregnancies ended in miscarriage each time. Every miscarriage brought with it such heartbreak and a new strong determination to succeed the next time.

My doctor suggested I go and see a fertility clinic and a specialist in this area. So off we went and had blood tests and ultrasounds and seems both of us (my husband and I) seemed to have every test known to man. Well that's what it felt like. I then was told I would need to have IVF.

I was given maximum doses of hormones allowable and was producing the grand total of five eggs, three of which would fertilise. However, because of the low quality of my eggs now that I was thirty-eight, the cells simply could not divide fast enough to support life in my embryos.

I never had enough embryos to freeze any, so each time I went back to try for another pregnancy, I had to do the whole IVF hormone-injection thing again. We did this three times, and I did not fall pregnant once. I did have one more natural pregnancy in between IVF cycles; however, this also ended in miscarriage.

Well, I was hooked by this stage. To come this far and give up? The yearning was worse than ever before. The doctors at the fertility clinic told me that my eggs were no longer viable. Depressed and inconsolably sad and bewildered at my nonexistent fertility after two years of "pregnancies" and hope, I gave up.

Then one day about a month later, I bumped into one of the nurses from the clinic. I was telling her that there was no hope. She suggested donor eggs. All my hope came rushing back. Really? There are people who will give you their eggs? Yes, she said, but being that in my country there is no financial gain for the egg donor and the whole IVF process is so invasive and difficult, our egg donors are far and few between; however, I could go on a waitlist for a donor.

I contacted the clinic, and they said yes, I could go on the waitlist but that was "years" long. I was already forty now. I didn't have years to wait and hope and dream. They said to me that there was another way if I was prepared to advertise for my own donor. I was more than willing to do this.

I ended up with one woman responding to donate anonymously. A couple of women who knew me asked if they could donate their eggs to me. Although grateful for their kindness and good intentions, I simply did not want a donor who might drop in to visit me and look at my child as if he/she were their own. Nor did I want my child to grow up with some dark secret that he/she was in fact my friend's child and nor did I want to hide something as huge as this from someone I knew I would love so much, i.e., my child. For me, love does not exist if deception is thrown in. Love and deception simply don't mix.

Of course presuming I would fall pregnant with the use of someone else's eggs, we went with the anonymous donor. I was so very very grateful to her (I still am). I know what she went through to donate her eggs, having been through the whole process three times myself. It was *huge* for her to do this for me and my husband and not even know us.

Should this have sent me warning bells about this industry? Maybe. However I was too far gone and lost in my yearnings. Even my husband gave me an out. He said that if we didn't get pregnant, it was ok. We still had my two older children, and we would just focus on them and we would have lots of holidays etc....live a self-indulgent life.

I was on a roll though, and deep down, we both still really wanted our own baby.

I received great quality eggs from my anonymous donor. I fell pregnant on the second attempt with an embryo from her egg and my husband's sperm. After another thirty-eight weeks, we had our baby.

Of course he was the most beautiful baby that was ever on this earth, and we were and are such proud parents (not excluding my older children)! Oh, I love this little guy so very much that there are no words to describe. Every part of him is cherished in exactly the same way as my biological children.

I just wanted to make this clear before I go on, because I am going to be completely honest in a way that many people call hypocritical and often produces angry comments and fear of having to acknowledge that children created via sperm/egg donation actually factually belong to another family's genetics and history. There is no way in which we can pretend these facts away. These children are who they are, and while we parent our donor-conceived children as we would our own and we love them no less.... Are we entitled to deliberately take them from their own biological families?

I started wondering about this.

I would look at my child and wonder how much of him was like my donor and her family. I wondered many things about his personality, his looks, the things he did.

I wondered when I walked down the street if I might be passing my donor—his biological mother (genetically and factually speaking). And don't get me wrong, I wished with all my heart he was completely mine—mind, body and soul.

I began questioning donor-conceived adult people who I would seek out and interrogate as to how they felt. Initially I was appalled at how ungrateful many sounded, that they were given so much love when there were so many unloved children in the world. Most of them were well educated, had loving homes and professed to love both parents who brought them up. However they also still yearned for their biological family. Mothers, fathers, grandparents, aunts, uncles, siblings, cousins etc....Some even had a couple hundred possible siblings because of sperm donation not being regulated in earlier days.

Well you know....what about my child? I was beginning to see some logical reasoning behind all this. I then checked out adoptees, and they felt the same about their families. The difference to adoptees was that usually they were adopted because of some catastrophic situations where their biological parents were unable to keep them.

Donated children don't have the luxury of an excuse like that. They are given away or sold in some countries before conception with the intention that they be created to live with another family. More often than not donated/sold anonymously by a donor who has absolutely no idea of what kind of home or parents or values this intended child is going to grow up with.

There are not anywhere near the same requirements to have someone else's baby through egg/sperm donation as there is to adopt someone else's child. Why is there not certain criteria in every country of the world as to the suitability of a child being placed in the care of the recipient parent? Of course I acknowledge that one of the donor-conceived person's parents is usually (but not always) a biological parent as well, though the recipient parent is not screened, and the fully-intended child is from a different family, in this respect.

Anyway a few years ago, we went on and found my anonymous donor. She wanted to meet us too. She fell in love with our son as soon as she laid eyes on him. She said he was special to her and she would love to know him as he grew and play some part in his life.

I did welcome this. Over a period of time, we met other members of his biological family. I remember clearly one day when our donor visited and she brought her mother along. I remember feelings of jealousy and envy surge within me with the realization that I was the only person in the room who was not biologically related to my son.

And there was so much evidence in front of me. He looked a lot like her. His fingers and toes are exactly like hers (which is the very first thing I noticed), and he moved like her.

I could not help but love her also, which I think stemmed from the total unconditional love that I feel for my son, and seeing them together brought such raw emotional feelings, because it was very hard to separate their likeness (them) and the knowledge that she in fact is his biological mother.

I had to keep my focus that I was doing this for my son's well-being, and this was hard to do at times like this, because I just wanted him to be mine. However I felt a strong responsibility towards him because it had been my "choice" that I take him from his bio family. I owed him this knowledge.

At the same time I am proud to be given the opportunity of being his mother. My older children are proud and happy to have a little brother whom they adore.

At this point in time, my donor had not had any children of her own. After meeting my son, she wanted children of her own. Yes I think strong emotional feelings were also stirring within her.

She went on to have (so far) three children of her own. Sadly her partner and her decided that they did not want their children to be brought up knowing my son as a half sibling. They told me that I was not to tell my son that their children were half siblings. I argued that my donor had wanted to be part of his life earlier and that I could not "untell" my son what he had already been given knowledge of. He had met a few members of her family and has an ongoing relationship with his biological grandmother. Even if I wanted to abide by their wishes, it would be impossible to undo what had already happened and what he already knew. For me, this was not an option anyway. My love for my son is unconditional; there is no room for deception and secrets in our relationship.

I may not be his biological mother, but I am his mother, and I will be someone he can trust to always act fiercely in what I believe to be in his best interests.

I am angry for him that he has been given away twice—once as an egg with intention of becoming a human being and again as a child. How do I explain to him why he is not wanted in their lives without causing him pain?

On the other hand, had I not grasped this opportunity of giving him knowledge of who he really is, would he have one day felt contempt at the deception of not having this information about himself? My answer is yes, this is what I believe. I have talked with many donor-conceived adults, and believe me, they know most of their lives that there was something amiss, and many feel a certain disconnection within their families, and much falls into place once they find out the truth. Had they had the opportunity to grow up knowing who they really are and having opportunity to form bonds with their own bio families, would this have made the difference for them?

One popular argument for the use of donor conception is that many children in this world need to cope with loss of biological family through all different kinds of circumstances, i.e., death, separation of parents, unwanted or unplanned pregnancies etc.…and these kids cope. Ask if this is ideal? No it isn't, *but* it mostly isn't planned either. In donor conception, this is planned.

I am finding now that I look at internet sites and there is a new age of recipient parents out there "making" it all o.k. in their minds to continue with this practice by accepting anonymous donors (eggs or sperm) and then registering their children, therefore doing the best they can to ensure that their children have access to biological family at age eighteen. Is this good enough? Their children may still never find a donor that doesn't update information or does not want to be found. Why are they finding it so important to attend to these registers? There must be a reason behind it. Could the reason be that they know that it should be the child's right to know bio family? Why then accept anonymous human reproductive parts? There seems something strange about accepting donor sperm or eggs, getting pregnant, creating a child, and then rushing out to make it all o.k. by registering the child to be able to find out about themselves at eighteen. This is too late to find out who you are. Teenage years especially are emotional roller coasters, and this is a lot for a donor-conceived person to deal with when information about themselves is denied. We do not ever "own" our children. They are people, humans in their own right.

Does all this sound complex and confusing and full of emotions and impossibilities? Well this is just one story of hundreds of thousands.

What are we doing to those we profess to love most? Of course, the majority of these kids do have very happy lives, so what can be so bad? Well....they mostly haven't had time for it to all register as yet, and for many, this may not even happen until they have children of their own and want to give their children their own unique family genetic information or a connection with blood lines. However, without family bonds and relationships in place, it is merely paperwork, and the loss is real.

Yes I can see exactly why we recipient parents are seen as making selfish decisions. As with my story, it all started with my feelings, my wants, and what lengths I would go to to get what I wanted. My infertility was not reparable, so I used someone else's. Whilst it got me what I wanted, it is wrong on so many levels. Not to mention the health risks. This is one thing I don't get with the whole fertility industry. They pump us full of what can be dangerous and life-threatening drugs, with the intention of bringing a child into our lives—or maybe an orphan, should these drugs end up being detrimental to our health. What is the point of that, and how ethical is this? Clearly the child being created here is not being put first and foremost.

It is my conclusion that the technology we have to create children is way, way too complex for us humans to deal with. Should we just accept our infertility, or should we continue to take children from other biological families and make them ours? In our own minds, that is—because the reality is that DNA and genetics cannot be changed, no matter how we choose to think. These children created by donors can never really be anything but another link in the donor's family—another generation of that family.

Now the above story is of course my thoughts and feelings and mostly "my story." With the loss we know that adoptees feel and the joy and tears and heartbreak of families finding each other after many years, the instant connection they feel when they meet a sibling and the deep sorrow they feel for not having had the opportunity to know one another, why do we continue to try and make it o.k. to do the same all over again—*intentionally*?

"Wanted Child"

I had never planned on being pregnant. I had always wanted a child but had planned on adoption. My husband had never planned on having children but then ended up married to his first wife and had a son. Having his son changed his life and his attitude of having children. He wished to have more. However, he got a vasectomy at his wife's request, and they divorced.

When my husband and I were dating, I was concerned about two things: our age difference (we are almost twenty years apart) and his view of adoption. My previous boyfriend of several years was against adoption. My husband was excited about having more children, whether through pregnancy or adoption, but he wanted them before he reached fifty. We talked and decided we wanted to adopt a child under three years old first and then adopt one–two older children after that. After looking into adoption, though, we were disheartened to find out the costs of agency adoption and the waitlist for social-services adoption. Oddly our insurance offered infertility assistance (but not adoption assistance).

I knew that people had conceived after vasectomies using IVF/ICSI. I was hesitant with going through IVF. I had never thought about being pregnant. After much discussion, we decided to try IVF/ICSI. If the first round failed, we would get on the waiting list for social-service adoption and try IVF again as well. While getting tested for IVF, we found that my husband's hormone levels were quite low. We were informed that in order to proceed with IVF/ICSI, my husband would have to have the sperm extracted by TESE (cutting open the testes and extracting tissue). Even with that invasive procedure, we were recommended to set up a donor-sperm backup. The doctors felt the chance of collecting viable sperm from my husband was around 20 percent.

Faced with this new information, we started to look at donor sperm. My husband was the one who preferred to use it. His family had some medical issues that he was concerned about passing onto our child. My initial response was "No." It seemed a little creepy, and I worried about how my husband would feel with a child that wasn't his biologically. But he reminded me that he was okay with adoption and this was just a different version of adoption.

We decided to forgo the IVF and simply try an IUI with donor sperm. For us, choosing a donor wasn't hard. Our reproductive clinic only worked with three sperm banks. We chose a donor based on physical traits similar to my husband and whose interests were similar to ours. The first donor we looked at we both liked. We reviewed others but in the end, went with the first one. He was a new donor, and his sperm had only become available six weeks prior to our order, so he had no confirmed pregnancies.

Before the IUI, our clinic required us to get a psychological consult, since we were using donor sperm. She asked questions about how we felt going into this (was it a choice or did we feel it was the only option), how our families felt, and would we be open with our child. We explained our desire to adopt and how we view this as a variation of adoption. We wanted to be open with our child and specifically chose a donor who was open. We told our families, who responded with happy, "We just want a baby" comments. We were allowed to proceed, and I am happy to say our IUI with donor sperm was successful.

Several weeks after my positive pregnancy test, I checked back with the sperm bank. In those few weeks, another positive pregnancy was reported. Currently I am nineteen weeks pregnant, and everything is going just fine. My husband is so giddy about the baby and loves telling people that we are expecting. We are both excited about the chance that at least one other person might conceive with the same donor sperm. We both hope the other parents will be willing to exchange information and possibility to allow the children to meet someday. I've already collected information to register with the donor forum from our sperm bank, as well as others that I've found online. We want our child to know about potential biological siblings. I am hoping to find a group locally of other folks who used donors to get their children, so that our child can see that they are not the only one.

Our child will get access to some sort of donor information when they turn eighteen. Both my husband and I hope they will be able to contact the donor. It is funny, as much as we are excited that our child gets to have this whole other world in their life, at times we forget how they were conceived. We find ourselves making comments and wondering, will they have dad's height, dad's big nose, or dad's charisma? We worry about how our child will handle this information. I believe it is normal to want to know about the people you come from. My husband and I knew with adoption that would be an issue to be addressed. But we figure this isn't much different, and we will do the best we can to support them.

We will let them grow up knowing the truth and let them make the choices they want in investigating siblings and the donor. We'll try to expose them to children who were conceived by donors or brought into a family by adoption. And most of all, we'll keep telling them how much we wanted them and how much we love them.

"Optimism and Confusion"

I am already a mom of one wonderful five-year-old son. I know I want another child. I've actually wanted another for about a year now. I'm strongly considering using donor sperm. I read through many stories of children wondering about their father. I never knew my real father. He left when I was three and my sister was one.

My son rarely (maybe once a year) sees his dad. I didn't make poor choices. We were married and both military. How was I to know it wouldn't work out and he would choose to go overseas? To leave the country without any regards to his firstborn. These are situations where the dad chooses to neglect the child *after* the child is born. I understand that most donors are choosing to ignore their children, but at least it's before they ever meet them. Before they should have fallen in love with them.

I want to bring another child into my loving home. I can more than afford another child financially and emotionally. At least my child won't feel the hurt I felt knowing that their dad met them and still chose to leave. One day I hope to meet someone who will love me and my children. Someone who could be a good role model to them. He doesn't have to be their "dad."

This was not intended to hurt anyone. It was just a justification for my choice of donor pregnancy. I've been planning, and I'll start trying in January.

"A Lifetime Yearning"

I was the child of a womanizer. My mother fell in love with a man who would never grow up, and I paid for it. My parent's marriage lasted only a year, and I always lived at my mother's family home, where I always felt strange and very much an outcast. Both of them were my biological father and mother, but I don't recognize myself in them. My dad was more an uncle than a dad, and he forced me to act as if we were living together and happy.

For years I waited that the promises he made would become true, because he brainwashed me with the fantasy that some time we would live as a family in a house of our own.

Since I can remember, I didn't want to get married, but to have a kid and to raise him or her with love, something I had not experienced, because my mother is a "dry" person. She did not receive love and did not know how to give it to me. I felt always more of an annoyance and thought that I owed my mum big time for the sacrifice she had made having me.

I grew up and fell in love, but always with abusive men. I had breast cancer when I was thirty-five and survived. I tried to get pregnant with a couple when I was forty, but I had two losses. Then I tried alone two times, with IVF and my own eggs, but they were already old. Three years ago, I had an IVF with eggs and sperm from donors. And a miracle happened, when I thought that the only thing I had really wanted in my life was already too far from my reach. A beautiful baby boy came in to my life. He is two-and-a-half years old, and I could not love him more. I know that what I did comes with a huge responsibility. I devote my life to him, and it is no sacrifice, it's what I always dreamed, to raise a wonderful human being. I know that adoption was an alternative, but in my country like in others, it is very difficult, and there is always the chance that the biological parent appears and takes away your heart.

I have given a lot of thought to the issue of how to explain to my kid where he came from. I think that I could have been much happier if I had not had a dad like mine. Sincerely, I would prefer to not have had any.

I am a teacher and see how the children suffer because of the indifference of the biological parents. The stories that some people share might have to do more with the fact that they didn't feel the love and care of the parents—were they biological or adoptive.

My boy is a happy kid who loves his mother, his grandmother, and recognizes his family. He knows I adore him, not only because I say it to him, but also because I demonstrate it in every way possible. I don't know if I will ever tell him the whole truth of his origin or only about the kind sperm donor who helped me to become a mum. I don't like lies; all my family, friends, and colleagues know about the IVF, but only my mum knows about the egg donor. It is really complicated, and I have come to believe that some truths are more harmful than helpful. What to tell, and when, is something that I had to think about, and I plan to go to a psychologist and ask for help. Maybe he will help me to find some answers.

Thank you for letting me share my story.

"Searching for Yehoshua"

Hello to you my dear friends.

My situation is like yours, but it's a little different. Some of you are children looking for your parents; well, I am a parent looking for my child. I want to encourage you by saying that your journey, pain, and searching all has meaning, and it's not in vain—even if you never find your biological mom or dad. Because you have love and yearning, there is meaning to your suffering. It's not in vain; neither is your searching and your sacrifice. I applaud you for having the courage to even come to this website and tell your story. You are special, unique, and worthy of love, especially a parent's love.

Three years ago my wife disappeared with my little baby boy, and I don't know how or where to reach him. It is anguishing for me not to hold my son and look in his eyes and be that loving, nurturing presence in his life. Since you the reader have wondered about your parent, if they would love you and care for you if they knew you, or if they could have something to impart to you, let me answer as a parent who desperately wants to be in their child's life. The answer is that deep down, parents want their children in a way that is deeper than mere words. Deeper than words, I say. There isn't a minute that goes by that I don't think of my child.

Even though you may not know it yet, your story has a redemptive purpose. I really believe that you can be a force for good in the world if you set a goal for yourself. The world needs people like you who can overcome suffering and let the world know that there is hope, even in the darkest times.

Yeshua (Jesus) came to earth to save you by becoming a forsaken child. "My G-d, My G-d, why hast thou forsaken me?" he said on the cross. You don't have to be fatherless in this world, because G-d wants to be your father. When you don't have hope and when you just feel like life is meaningless, you can always come back to Yeshua, knowing that G-d loves you.

"We Have Children..."

I married young and have genetic and biological children. They are a blessing. That marriage ended, and years later I met a wonderful man. We were older and knew my ovaries were unlikely to do the job, but he came to me without any children of his own, and he wanted kids, and I love kids, and so we decided to try to have kids.

I'm blessed to be the biological mother of our children. The genetic bit, the blueprint if you will, came from another woman who donated. I'm o.k. with that. I wanted *his* kids to love, and I got to gestate our children and so have that biological connection. The genetic link isn't irrelevant, but it also doesn't matter to me, since I got to pass on my genes already. The gestational influence is pretty substantial, since all the "switches" to be thrown were done by my uterine environment.

When they are older, they can seek out contact with their anonymous donor, and I hope they are pleased with what they find.

Meanwhile, I'm blessed to have our children.

"I Worry About My Daughter"

Don't think as a parent I don't worry or realize what I have done. I have brought my very much wanted and loved daughter into the world not knowing who her father is, and that will be frustrating for her. It puzzles me at times, cause I am sure some of her little unique traits are from her donor daddy. I know my parents, siblings, grandparents, and great-grandparents, so I can't begin to understand how she will feel not knowing all of hers.

She is a unique child in our family. The only grandchild and very much adored. She was born to a Choice Mom. But as I had the choice of having a child now or quite possibly never having one, I chose to do so with a donor. I know his donor number, his nationalities, hair color, hat size, etc. I know what he is allergic to, who he resembles in famous circles. I have his medical history up to the time he donated, and his parents and grandparents as well. We are in touch with nine of my daughter's siblings, thanks to the Internet and donor posting boards. I will register her with the Donor Registry and am very open to knowing her donor father if he ever wants contact. Do I wish that I had found an open donor? Or could afford to have tried with sperm from a bank that gives out identity when she is eighteen? Sure....but the costs were beyond my bank account.

I hope knowing about and hopefully in person her donor siblings at some point will help her filling blanks and connect not only with other kids who have a donor daddy but are one-half biologically related to her. Knowing other kids who have a donor daddy I think will be key in her identity. I hope to connect with other Choice moms as well. I plan to do everything I can to answer her questions. But no, I will never know exactly how she feels. I am guessing from reading postings she will always feel a missing part of her identity. But perhaps not....just like some adopted kids don't feel the need to find their bio parents.

As a Choice Mom, it is no secret as to how she came to be nor will it ever be. Do I hope to have a Social Dad for her? Sure...but with my luck dating, it is still only a hope.

I plan to share fairy tales with her about other types of families. We never know who Snow White's parents are, or who Cinderella's dad is. The Little Mermaid and Belle from Beauty and the Beast didn't come from traditional families either. I hope that knowing her bio siblings will help. I plan to ask her how she feels about being part of a unique family and listen when she probably will get mad at me about it. But I will do my best to be an amazing mom and give her all I can.

Do I wish we were a two-parent household and she was part of someone I love deeply? Absolutely....I know that for lots of reasons, those types of households are the best. Do I think ours will be a bad place to grow up? Absolutely not....Life holds no guarantees.

I had friends growing up from divorced families, blended families, one where three girls were being raised by their dad and mom was someone they only saw once a year, and....One friend in particular never had a dad that I knew of, and she was biracial with a white mom. No one ever gave her a hard time about it. She didn't have a dad....So what, we didn't care as kids. We all knew he had to be black....but it was really a nonissue. I hope to surround my daughter with people and other kids like that. Ones that don't really care she doesn't have a dad at home.

I just hope each of the donor kids out there know that the donor had a reason for helping us have you…and we really wanted to be your parent, and am glad he allowed us to do so.

"We Used a Donor and a Surrogate to Have Twins"

My husband and I met and married late; I was in my forties, and he was fourteen years younger. We wanted children, and I got pregnant right away but had a miscarriage. Nothing happened after that, so we consulted various fertility specialists and went through various procedures and mishaps. After the fifth specialist, we gave up hope of having our own child. By then, I had developed another medical condition, which made it impossible to even carry a child.

After I met a friend who adopted a child, we decided to do the same thing and visited two adoption agencies. I also used the Internet to search for more adoption info. It was while looking for adoption info that I came across egg donation and surrogacy. Before that, what we knew about surrogacy was related to the "Baby M" case, i.e., the surrogate mother decided to keep the child she was carrying. Since the baby was also her genetic offspring, the courts granted her visitation rights. We did not want to risk this, so ruled out surrogacy. By using an egg donor as well as surrogate, we would not run this risk.

I also was concerned that our future child might resent not knowing about his/her genetic background, so I made sure that the egg donor and surrogate would be open to contact.

We were fortunate. We have beautiful and healthy twins. We've kept in touch with the donor and the surrogate and set up a website where they and the twins' grandparents can see photos and get updates. The kids met their donor last year, but they were too little to remember, so we took lots of photos. When they get older, we plan to take them to meet their surrogate as well as their donor again and resolve any questions they might have.

Finally, I'd like to address the issue of egg and sperm donors giving their "babies" away by donating ova and sperm. An ovum and a sperm do not equal a human embryo or a baby. They are needed to make babies as are carriers or surrogates, but they in themselves are not babies.

"Thankful for My Donor"

When I was thirty-seven, I discovered I would have a very difficult time getting pregnant because I had prematurely aging ovaries. At the time, I did not have a partner. I always knew I wished to have a child. It was then, and especially after my insurance immediately agreed to cover the expensive IVF procedure(s), that I knew it was now or never. I myself am an only child. I do not come from a large family. I decided to have a baby on my own, because I knew I would be a great mother and that I wanted more than anything to have a child. But this decision was not simple. It took me a long time to come to terms with it. One thing I took into account was the role my own father had played in my life.

My father was always quite critical and even stated on occasion that he'd never wished to have children. This devastated me. So, you could say, I did not see the fact my daughter would *not* have a father as an entirely bad thing.

Having my daughter (now two-and-a-half) was the best decision I ever made.

When I was pregnant (after three IVF cycles), a man asked me what I would tell my child. How would I explain myself? (This from a man who'd fathered more than one child out of wedlock.) I said this: "I will tell her the *truth*."

And I will. Because she deserves it. If and when she wishes to find her bio dad, I will help her. In the meantime, there is no one I love better than my daughter, whose existence was hard won. It took a lot of work for me to get pregnant. And I'm grateful for the donor who made that possible.

My daughter is one of the happiest, funniest people I've ever known. The world is a better place with her in it. And I intend to always make sure she *knows that.*

To all those women out there who think they have to wait for the right man to come around in order to have a kid: *You do not have to wait.* Single motherhood is hard, yes, but also totally awesome. My daughter and I are a team. Her grandparents (even my supercritical father) adore her. Simply put, she and I have a great life together.

"The Best Policy"

I am the coparent of a boy, now three, who was born via IVF using American donor anonymous sperm. My female partner gave birth to him, and we live in Australia.

It has always been our strong belief that we will tell our son everything true about his origins as early as we can—although navigating that situation is tricky!

Like most parenting, it's a case of taking it a day at a time and using your best judgment while fervently hoping for no lasting psychological or physical damage! Seriously though, we have already started telling him how lucky he is to have two mums, how some other kids have two dads, some have just one mum, some only a grandpa, etc. Once or twice so far, I have told him that we needed some magic ingredients from a donor to help create him, but the concept of reproduction is a little outside the boundaries of a three-year-old's understanding, I believe. So right now, we keep it simple.

We will keep explaining it though and keep making the effort to find the right words, because it is so crucial. Why do we feel it is important?

Because it goes to the heart of the identity of our son. Not just what he looks like, or his blood type, or why his hair curls, but his character traits, interests, abilities, and personality. Nature *and* nurture make a human.... Each to varying degrees, but each must be respected.

Our job as parents is to nurture, protect, inform, educate, discipline, and observe the formation of that identity—what a scary privilege, and I wouldn't dare consider withholding any truth that might negatively impact on that.

It has been interesting to read some stories from the U.S. and the laws relating to tracking down donors. Our IVF clinic has told us that when our boy is eighteen, he can make inquiries about his donor in the U.S. and most likely find him.

We have already made the decision that we will support him, whatever he decides on that front. So, the term "anonymous" donor is applied loosely here in Australia. And we knew that going into this, so we accept it.

We also had some counseling before we conceived, through our IVF clinic, and the counselor took my breath away when he said, "Do not think of the donor as a father. Your child will not have a father; he will have two mums."

Until that point, I had been scared about applying the term "father" to this man, because we knew he would play no part in raising our boy day to day.... But hearing that perspective really helped.

Of course he was important and essential and a human being full of wonderful faults and talents—we will never forget that. But my partner and I have raised him. So it's an endless balance of both truths—and making sure he is as informed as he can be. Knowledge=security.

We are in the midst of trying to have our second baby.

"Possible Sources of Children's Discomfort"

I am the father of two adopted children. I know this website is supposed to be relevant to children conceived by donor sperm or eggs. Yet, I think what I have to say may be important.

When we started the process of adopting our first child, the adoption agency strongly advised that we tell our new baby that s/he is adopted. True, newborn children cannot process this information. But we were told that by the time they can process it we, the parents, will feel very comfortable in communicating this information. It will all seem perfectly normal and natural to the adopted child.

We followed the instructions faithfully. When we went through the process of adopting our second child, the first was made to feel that he was a contributor to the adoption process.

Then when our children were teenagers, we told them that if they would like to know about their biological parents, we would help them find out. At first they were not interested. As they got older, we repeated our offer to help find out about their biological parents, and they remained uninterested. We do not know why this was the case. Perhaps because being adopted seemed like such a natural event and they both knew quite a few other children who were adopted contributed to their disinterest in knowing about the strangers who had only one short contribution (we never used these words to our children) to their existence.

If parents would be up-front, from the beginning of their children's life, about how they came into this world, maybe children born by donor sperm or eggs would grow up much more at ease with their lives. Some will want to know about the egg or sperm donors; others will not care.

I sincerely believe the advice the first adoption agency gave us was right on the mark. Physicians who perform IVF should be required, if they are not already so required, to advise their patients to communicate the facts of the birth of their children as soon as they are born.

"Choosing a Known Donor, a Friend"

After ten years of marriage and many years of waiting and hoping for pregnancy, my husband was given the diagnosis of infertility. His sperm count was zero; there was no chance whatsoever of us producing his biological child. Initially we decided to remain childless, but I fell into depression, an illness I have never faced before or since then, but it was horrible. And the issue had in general put our marriage at risk. So we sat down together one evening and talked through all the options. Within a short amount of time, we were in agreement to try to become parents with the help of a sperm donor. We chose a man who we had known for several years. Someone we knew, respected, and who was in a stable marriage with a son of his own. Why did we go this route? We wanted our child to have the option to have a relationship with the biological father. Picking someone we knew and liked would make this easier. And also as a mother, the idea of becoming pregnant with sperm of a "stranger" was very, very unattractive to me.

Fast forward ten years. We have a healthy, happy third grader. Our child knows their biological father and is at ease with the situation. Because we live in different countries, the two families get together infrequently, but when we do, it is just like any other "family" visit with both the good and the wonderful.

Although I haven't seen a lot of talk in the Internet about choosing a donor from your own group of friends, I suspect it is not uncommon. Done properly with a mature and honest attitude, it can be a simple answer to a very difficult question. As one of my closest friends said when I asked her to give me her opinion about our crazy plan, "No child was ever hurt by having too many people love it."

"A Positive Story—for a Change"

I heard about this website from a link from a friend's blog. My interest is that I am the mother of a child conceived by DI. My husband was unable to genetically father a child, but it does not stop him from being the best Daddy ever to our daughter.

The reason why I felt the need to post is that it truly bothers me to see some of the posts on here about how parents make unethical choices to conceive a child via donor. I would have been more than willing to adopt, but because of my husband's medical issues, we are unable to. Does that mean that we are unfit to parent? I hope that we will not raise a person who will make snap statements about people's life choices without knowing the entire story. And is this the ideal situation to bring a child into the world? Of course not! Not as long as people are as small-minded as they are. And as to why I would want to conceive a child with a stranger, I didn't. I conceived with my best friend holding my hand and telling me he could hardly wait to be a daddy.

Our daughter is smart, loving, creative, and beautiful, and my husband and I thank G-d for her every day. I am sure we will have issues about many things in her life—maybe donor related—but I will not worry about that now, just as I am not going to worry about other problems that may or may not happen. The best thing we can do as parents is love her unconditionally and be active in her life.

Deciding to use a donor was not an easy decision, and the process is emotionally and financially difficult. But it is easy to be a mom and a dad. (Not that I am saying parenting is easy—but it is just a natural part of our lives right now.)

I am not saying that this is the right answer for everyone. Fortunately, for us it works, and we are beyond happy with our decision. I have no regrets—"regret" is not a word that I use with my daughter.

"DI Is Never a Good Option"

My husband is infertile. We chose not to conceive by DI, and I am thankful we made that choice. The world is already filled with children and adults who struggle daily because of their parents' unethical choices, so why add more kids to the world who will suffer? Conceiving by DI takes a lot of financial and emotional resources and a lot of energy that could better be used to help the already existing children of the world who need help.

I can't even begin to understand why somebody would want to conceive a child with a stranger. It is mind-boggling.

"It's Not a Perfect World, but It's My World"

Who could've guessed all the emotions that a parent could have toward a child. When I was young and learning that having a child, adopting a child, even creating a child within was going to be an almost impossible task, I kept the positive attitude that "I can do this." And it became so. I am now the mom to two beautiful girls that are loved to the ends of the earth. That's not to say it hasn't been without its ups and downs. But for me, the ups are far, far more than the downs. And I know the emotions about one adopted child and one donor's child's genetic beginnings are deep and wrought with questions about ethnic heritage. Where is the family that originally gave the adopted child up? Where is this man that donated half of the donor child? Why can't she find anything out about him? Are there any genetic siblings out there who also want to know my girls? And I can't help, because I've given them all the information that I have, and it all seems to lead to nowhere.

In hindsight, I should've demanded more information of the donor father. I even might have used a known donor vs. an anonymous donor. At the time, I didn't think that would be the moral thing to do. As I've aged and perhaps grown in wisdom a bit more, perhaps that would've been the better thing to do. Or perhaps I should've never married an infertile man knowing how badly I wanted children. Can't go back now. Can't get a redo on that chapter of my life. Biggest mistake made—marrying someone who couldn't produce and didn't want children. And then complicating the whole situation by adding an adopted child and a donor child to the whole mix. Adding the children wasn't the mistake. It was the husband that was the mistake. The children have been the greatest joy of my life. I'd do everything I did to add them to my life again in a heartbeat. They always have been and always will be the most important part of me. Rain or shine. Bad or good. Can't imagine my life without them. The wisdom they've afforded me has been immeasurable.

Guessing by reading these articles that everyone out there has issues of some sort or the other. Regrets, yes. Would like to redo, yes. But we can't. We do what we can with what we have—no matter the circumstance. Some people do a better job of moving forward to make life better than others. But my wish for all the people in turmoil out there is to find peace. Find a way to enjoy what you do know. Find a way to resolve the unknown in your heart and enjoy the known. Take the relationships that you are blessed with, work through the troubled spots, and keep them strong.

"I Could Not Be a Donor..."

I am sterile. I told my girlfriend when we met that I could not make babies. She accepted that and said we'd deal with it when the time came. A few months later, we were married. We had fun, enjoyed life, and played lots. So when we were ready for the next step, we investigated adoption and fostering. We went through the paperwork, home studies, investigations, and payment programs. After a few months, the adoption agency presented a picture of a baby girl. She was beautiful, and we said we would definitely accept her. We went home, made all the arrangements, decided on her name, bought all the required items, and waited. Our baby was one month old when they showed us the picture. She was six months old when we got her. I was the first to hold her in my arms, and I was truly blessed. She was warm and beautiful and sweet. We showered her with our love and spirit and did everything we could to help her grow. We were twenty-two when she arrived.

After about three years, my wife wanted another child. I thought one was plenty of work, but her body was telling her it wanted to birth a child. So we started investigating artificial insemination. Oh, I did go through the barrage of tests to see if there was one chance in a million that I could be the father. I wore boxers to keep the gems cool, she did the rhythm program and taking her temperature on a daily basis, we did everything we could to get a donation from me...but to no avail. I had, as the doctor so gently put it, one live sperm per ten million parts, and that was even inconsistent. Yes, I had to make more than one donation to the cause. My wife had to take drugs that flushed out nodes in her womb and they also messed with her mental balance, so she was very difficult to live with sometimes. Which is when we began arguing about whether having another child was worth all this. It was a struggle. But we kept going. We prayed about it, and so, when she went back to her doctor, he said there was a program at a nearby hospital where they were looking for sterile couples to participate in a study and possibly get artificial insemination via a donor clinic.

My wife and I talked about it. Three things were important: The donor needed to be educated—have at least a bachelors degree; needed to be similar in appearance as me—fair hair and medium build; and blue eyes—like me. I was all for it, because I just wanted to get past this point where there was nothing else on my wife's mind than having a baby naturally. But I wasn't perfect. I had other interests and so from that decision on, since I didn't need to be involved in the conception, I did other things and let her make all the trips to see the doctor. I took care of our beautiful adopted baby girl. We played, we sang, we danced. And one day, my wife came home and said, "It took! I know I'm pregnant!" So we all rejoiced together, because my wife was happy.

She did everything right. She didn't drink or smoke anyway, so that was the easy part. I did plenty of drinking to cover that. I was an idiot in my twenties. And so….nine months later came this beautiful little fair-haired girl. Her older sister's initials were similar to her mothers, and hers were the same as mine. We had our namesakes; her mother had two pretty and intelligent daughters who would always be there to take care of her in her aging years.

We had told our older daughter that she was adopted when she was about five years old. And continued on as she was growing, because there were other adoptive parents around with children she played with. But the deal on telling the younger daughter about her beginnings was another subject completely, and I really don't remember my wife ever bringing up the subject of telling the younger about it. So, when my wife and I did finally go our separate ways when the girls where fifteen and ten, and my wife suggested I relinquish custody (or she would go for more child support), I talked to the girls and said that their mother needed them and that she would always make sure they had nice clothes, good schools, and all the things they would need to get where they needed, and if they ever wanted to see me, all they had to do was come by. It wasn't that I didn't love them; it was accepting that their mother wasn't going to be happy as long as I was still a part of their lives. So I signed them away and have regretted it every day. I want to do whatever I can to support my daughter and her interests….I still love her, even though I can't be a direct influence—she was brought into this world to have a life, to make a difference in other peoples' lives and to take joy and pride in herself! I am very proud that I was a part of it. I also know there are many children who don't know who their donor parent is—via rape, promiscuity, and adoption—but they lead their lives without knowing that history. And even the ones who know their history, but totally ignore it, are continuing to do good things just for doing good things. Thank you for this opportunity!

A Sterile Man

"Proud to Say"

I am proud that I said "No" to conceiving a child with a man who donates sperm. Fifteen years ago, my husband was diagnosed with male infertility. We chose not to conceive a child with donor sperm for ethical reasons. It was the best decision we ever made. We decided to adopt two children.

I have never been pregnant and never will be, but if I had the privilege of conceiving a child, I couldn't imagine mixing my blood and genes with that of a stranger to produce a new human being. I couldn't handle the idea of a child growing inside of my body, and half of that child's identity belonging to a stranger. It is a huge leap from what I always planned for my biological child, which was a child that was biologically half my husband and half me.

I am not very good at handling guilt. If I had conceived a child with donor sperm, than I would feel terribly guilty for placing the burden of my husband's infertility on my child. My child would pay a huge price in dealing with the challenges of an unethical conception. It is a parent's job to make sacrifices for their children, and not the other way around.

I talked to a couple of married women who had conceived a child by donor sperm. They both said that they were very emotional and scared during their pregnancies, because they conceived with donor sperm. This is definitely not the type of pregnancy I want. One of the women kept her daughter's identity a secret from her, yet she was willing to tell me about it. This was really sad; the mother told a stranger the truth but would not tell her own child.

I also believe that conceiving a child with a man that is not my husband is adultery even though no sex is involved. This is true whether the donor is anonymous or known. By conceiving a child with a man, I am creating a special intimate relationship with him, even if he is anonymous. I would naturally have feelings of closeness with a man who gave me the privilege of having a biological child.

I mostly used the term donor sperm for convenience, but I think a more accurate way to describe this is to call it "conceiving a child with a man who donates sperm." Saying "donor sperm" depersonalizes it and makes it easier to digest, because it is about conceiving a child with an "it" and not a person. The reality is that people conceive children, not objects.

No regrets.

"Why We Chose Donor Egg"

My husband and I have a biological daughter who is seven and have been trying for five years to give her a sibling. We are both on the older side and for whatever reason, were not able to conceive again the old-fashioned way. We tried IVF four times with my eggs and his sperm. No luck. So we tried donor egg, and it worked. I am now about to deliver our second child—one who is genetically related to my husband but not to me.

We are aware that this child may feel anger, resentment, and loss at not having been raised by his biological mother, but we will be honest and open about his origins from the beginning. We also made sure that the donor would agree to be contacted by this child when and if he desires.

I want him to have that ability—to know who she is and to be able to contact her. I agree that the situation is not ideal—two biological parents is probably ideal—but many, many children never have that, and many who have two biological parents are in terrible situations.

Our son will always be loved and safe and cared for. He will always know where he came from and why he came into this world in an unusual way. I have never believed in the myth of some picture-perfect, mainstream, cookie-cutter family life. Life is not perfect. No one gets to choose who their parents will be. A strong, loving family is rooted in much more than biology.

I also know that this child has literally been built from my body….His every cell has grown from my blood. Every choice that I made while pregnant—what I ate, when I rested, how I played—had an effect on his development, helped to activate or deactivate certain genetic qualities. This dynamic is called "epigenetics," and it means that my choices while pregnant influenced how my son's genes will be expressed. So, to some small degree, I have influenced him biologically.

All that being said, after having read the stories by donor-conceived people on this site, I am even more determined to be honest with our child. And I know we may be in for some rough times as he comes to terms with his origins. In the end, I hope coming to grips with who he really is will make him a stronger person.

We feel incredibly lucky to soon be welcoming him into our lives.

"A Real Dad Speaks"

In 1971, when I was about two years old, my parents asked me to name my rocking horse. It was the first toy I got to name for myself, and I chose "XYZ." Not long after that I was reading, and I continued throughout my childhood to be recognized by adults as being unusually bright. I won writing competitions and math competitions through high school, and I scored safely within the ninety-ninth percentile on both the math and verbal sections of the SAT. I was the one kid from my public school class of about 600 who got into Harvard.

I don't report all of the above to brag; this is an anonymous forum, after all, so you do not know who I am anyway. I'll get to the point. Fast forward a few decades, and I'm forty-one now, with a wife and a four-year-old son who is our world. When my son was about two, my wife and I asked him to name his stuffed turtle; until then, we had always called all his stuffed toys simply by the names of the animals they resembled. He knew those names and, an early talker like his dad, he could say them clearly. Yet he chose the letter "F" as his turtle's name. During my next conversation with my mom, I told her about that curious choice, at which time she reminded me of my long forgotten (but as soon as she said it, eerily remembered) XYZ.

Aside from the eerie coincidence, my wife kids me about how much my son is like me in almost every way—from his natural talent for music coupled with an inability to sit down and practice it, to his fine motor skills that somehow serve him well for any purpose but sports, to his penchant for wordplay. For example, when he was three, after he remarked that an apple slice I was about to eat looked like the crescent moon, I playfully downed the slice in one bite and asked him what it looked like now. Without missing a beat, he blurted out "the new moon," but then he quipped, "No, the new apple!" He had me at new moon, but when he took that a step farther than I had set it up to go, I knew this was version 2.0 of me. Recently, at the urging of his preschool, whose staff had been telling us for a while that he was unusually gifted, we had his IQ tested in contemplation of applying to gifted education programs. Sure enough, his IQ came in at a little over 150—99.9th percentile. Like father, like son, I guess, except that it seems he is a little smarter than I am.

My son is donor conceived, because I had cancer as a kid and have no sperm of my own. So why do I click with him so? Shouldn't he seem like someone else's child, somehow? From the paperwork we have on the donor, that man does not seem to have been as bright academically as my son and I are, and he does seem to have devoted himself rather intensely to music and film—while my son can be satisfied with ten minutes a day of either of those things, so long as he is given ample time to draw, read, do puzzles, and talk my wife's and my ears off. Isn't it possible that my son and I click so well because I took three months of paternity leave after my wife went back to work, cuddling him and paying attention to his early cooing more than many dads are lucky enough to have time to do? Because I changed half the diapers, and got up at 3 a.m. half the time when he needed that? Because I still read to him every day, and sit down on the floor to play with him whenever I can?

I read a lot of blogs about donor conception, many of which are written by adults who were conceived with the help of a donor. Most of these bloggers are so unhappy with the circumstances of their birth that they are motivated to write blogs frequently, and often to advocate elsewhere, for change in the fertility industry to address their concerns. A very common complaint is that prospective parents considering using donors—whether it be single moms, gay couples, or heterosexual couples with a fertility problem—fail to consider their prospective children's rights to know their biological parents. The laments in these blogs often include an attempt to refute the responsive argument that without assisted reproduction, these people would not have been given the gift of life at all, and would not be here to complain about the circumstances of their conception. Some such attempts go so far as to suggest that being donor conceived is worse than not being born at all. Others stop short of that, but all seem to agree that it is selfish of prospective parents to use donors, because the resulting offspring likely would not choose the circumstance of being donor conceived.

My answer to that is that yes, it is selfish of parents not to consider the viewpoint of the unborn child. But all parents considering starting a family are thinking in large part about themselves, and no child, however conceived, gets any say in the matter. That goes for children born to wealthy parents and poor, to drug addicts and abstainers, to loving or abusive parents, to single parents, to parents who already have too many children, to parents who have no intention of providing siblings, to parents related to a quirky Aunt Ida who smells bad. Unborn children simply do not get a chance to approve applications from their prospective parents.

I respect the writers of these blogs, as they have genuine feelings and they address them with passion. But with all due respect to their feelings, it is narrow of them to condemn for everyone what they have decided did not work for them. My son is not doomed to share their existential depression simply because he shares with them the fact of being donor conceived. Indeed, many of these bloggers have a beef with more than just the use of donor conception. Not all, I know—but many. Some are bitter because they were deceived by their parents about their origin, while others feel deprived by the absence of a second parent. But yes, some may have been raised without those additional problems and still wallow in misery.

I love my son. He already knows about his origin, and although at four he is not that interested yet, he will always have known his story and he will never feel deceived. Also, he and I are very close and loving, so he will never feel he was missing a father figure. Even so, I cannot guarantee that his circumstances will never plague him as those circumstances plague some. Yet I refuse to believe that the loudest, most vociferous opponents of donor conception speak for the whole class. Instead, I think that the ones who are not so troubled simply are not so visible. After all, why would they devote themselves to blogging about being at peace with their origin? I hope my son never develops the intense, existential pain I've read too much about, and if he does, I will do whatever I can to be there for him, as his dad. But what I wish for him is that he will grow up just like I did, and just like lots of people do: glad to be alive, and at peace with himself, his family, and the world around him.

I do not apologize for choosing to bring this wonderful person into the world, and I would certainly do it again.

"Chaos of Emotions"

Twenty-five years have elapsed since that very first moment I held my daughter in my hands. From the very second that I began warming the donor sperm from the cryogenic container to my body temperature I knew that my child could possibly be inside that tiny vial. And my deep love began for her precisely then. The contents of that vial did create an embryo, and that embryo grew to be a baby girl. I nurtured her inside my body. I coaxed her out of my body and into the world. I nurtured her in her infancy. I nurtured her through puberty and the teenage years. I pushed her through college and let her go into the world. And I loved every moment of it and every moment that I spent with her. The parental love was and is strong and did not nor will not change through the years. She is my child, and I am very proud and happy to be her mom.

But the tortures of a donor child are different than other prepubescent children, teenagers, and young adults. I dealt with those tortures in addition to the normal ones that a typically biological child would deal with. Being called a "test tube" baby in the days when the technology was new and not well understood was torture for a middle-school child. Searching for a father so you can understand why you do things in certain ways knowing that genetics plays an important role in the "code" of who you are is torture for the young adult. Feeling like you don't belong to the father that raised you because your "code" isn't anything even close to his "code" becomes a daily downer in your emotional well-being. The question has arisen lately about why I would even consider bringing a child into the world where I knew she would probably never meet her biological father. And my question back is, "Would you rather not exist at all than to exist as you are?" Yes, I would do it all over again and bring you into the world, given the same set of circumstances that existed then. I cannot imagine life without my beautiful daughter.

If I could have changed the set of circumstances that existed, then I would've loved to have produced a typical biological child with the same genetic code of a mother and a father living in the same household. I spent a lot of time in doctor's offices, taking tests, undergoing invasive procedures, taking drug treatments, and passing psychological exams trying to make that happen. But I wasn't given that opportunity at that point in my life, and I wanted that baby girl so badly that I took her just as she was with all the social complications that a donor child would bring.

And just so you know, I wanted children so badly that I worked at it for years. I have one of each as I like to say. Oprah should invite me to her show to talk about all that I've experienced through the years with the three children that are mine. I have an adopted child (doesn't know either of her biological parents, raised by an adoptive mother and father) that took me three years to bring into my home. I have a donor child (raised by me, her biological mother, and an adoptive father) that took five years from beginning of quest to birth. And I have a young biological son (raised by me and his biological father) that came along fourteen years later. All three children are considered brothers and sisters in our eyes and in the legal world, too. But the differences and the issues are all very complicated and all very different for each of the children. Each one is a different story for a different book. Too much for 750 words or less.

But now, twenty-five years later, the rose-colored glasses are off. Do I still think a mother's love should be enough? I do. But my daughter doesn't. She thinks she needs that biological father's love, too. And that hurts. It hurts me, and it hurts her adoptive father. But we move on and know that we're not the first parents whose children have questioned their decisions and their ways before. You can't allow yourself to be tormented by a child's questioning nature. My strength comes from within, and my love for my daughter will still be there in the end. Baby girl, I love you.

17. Personal Stories from Partners, Friends, and Others

"What It Feels to Be the Partner of a Sperm Donor"

I am the wife-to-be of a known sperm donor. My partner donated to his lesbian good friends twice before he met me. One of them became pregnant, and a baby was born who died aged four months. They asked my partner to donate again and he did—about six weeks before he met me.

I found it very difficult to accept that there was another woman out there carrying my partner's child. I did not want anything to do with the arrangement—as a mother of two children myself, I could not understand how he could contract away his rights to a child, or how the nonbiological mother was able to acquire a child in that way. I felt jealous that he had made this woman pregnant. It made me feel as though I was not in an exclusive relationship with my partner. I never accepted it; I refused to have any dealings with them, and my partner has accepted me on the basis that I find it really difficult.

I know that the lesbian couple wants more children by my partner and that my partner won't do it again because I find it so difficult. He was in a difficult space when the lesbian couple approached—I think he did not think he would ever repartner or want more kids—so his decision to donate was made under different circumstances to those under which he currently lives. He has a relationship with the lesbians, but I know it's not as free or easy as would otherwise be the case without me around. I can't understand how he can accept viewing pictures of his daughter on their Facebook site as they randomly pop up, without wanting a proper involvement in her life.

Friends of mine think I am being ridiculous, but I do! Sperm donation affects so many people, and I think sperm donors should consider the consequences for them and their future, too. I don't know how to tell my children that their half brother has in fact got another half sister who is nothing to do with them. I don't know how to face the lesbians. I don't know why I find it so hard. My partner donated for the best reason—he just thought he was giving nice people "the gift of life." I think he gave away his child. It's a bit like the Emperor's clothes....This child is not the product of that lesbian couple, and a father should count for something....But people are too scared to say so for fear of not being politically correct.

I guess I have to encourage my partner's relationship with the little girl he fathered in this way. It's not her fault. I know all this. But it is hard.

"History and Genes"

I'm not going to make judgments here—just an observation or two.

There have been "nonmarital events" for a long time. Things like women bearing children other than those of the men they were married to. And unwed mothers. Orphanages were once commonplace in Western Europe (Dickens's London comes to mind). And nobody rational ever suggested that all "legitimate" children had caring, decent ("legitimate") parents. Certainly, there was child abuse and abandonment, even incest and rape.

But in all this, the concept of families and some kind of "history" was part of society, even if great-grandfather was himself adopted. Every ancestral history begins someplace in time, be it in 1400, 1900, or 2000. Genealogical knowledge is often— perhaps usually— abused by social climbers and snobs, even (rather ironically) in the United States.

While I would never suggest that anybody "needs" to know his ancestry in any line over ten generations (more than a million ancestors at that point), it might be interesting to know that a great-grandfather served in the First World War, for example. No, we should not take credit for the deeds or misdeeds of our ancestors, none of whom was perfect. Truth be told, some ancestors are probably best forgotten. It's about the knowledge.

But don't we have a right to at least know something about family history, which we can then decide to embrace or reject?

Granted, adoptees rarely learn much about their biological parents, but that has sometimes prompted personal quests for information.

We may not like what we learn, and in any case, it may not be especially "instructive." The whole ethnic identity thing, for instance, is misplaced as often as not (the places we now call Italy and Germany didn't even exist as nation states until the middle of the nineteenth century). Here I'm talking about basic, simple ancestral identity, as well as medical histories.

Genetic genealogy, with its focus on haplogroups and ancient history, fascinating though it may be, cannot address the desire for essential information regarding the last few generations.

No, I'm not part of the family-history bandwagon. It's just that kids ask questions. They're curious. Establishing a relationship of sorts with a biological father (sperm donor) may not be sufficiently fulfilling for everybody, though I agree it's a good starting point.

Part of the "problem" is that the child of, for example, a single woman who chose artificial insemination may understandably feel the need to defend his mother and her choice—even though the decision itself was made before the child was born or even conceived.

True, we create most of our personal identity, but we shouldn't presume to decide for others what they may choose to identify with in their ancestral histories. Some of it might just be worth preserving.

"Perspective"

When two women claim they are capable of having the same relationship as a man and woman do, the implication is that every man's contribution to relationships, marriage, and family is replaceable and unnecessary. That is what is *really* a "war on men." And two men claiming their relationship is just as much a relationship or marriage as a heterosexual committed relationship implicitly dismisses any contribution of every woman, thus also claiming women are inconsequential to relationships, marriage, and family. That is what is *really* a "war on women." And both are *real* (gender) discrimination.

Also, comparing same-sex "parenting" to single parents is extremely faulty, because (1) people don't typically get divorced on purpose and (2) there is always a chance that an opposite-sex parent can come into the mix and help raise the child. With same-sex "parenting," the child for the lifetime of the relationship and *by design* of same-sex relationships is always deprived of either a mother or a father. It takes more than just two people loving each other to give the best environment and outcome for a child. Men and women are complementary opposites psychologically. And parenting is not about what the parents want or about using children as "equality" trophies to justify or validate a particular private bedroom habit.... Parenting is about what children ultimately need. Therefore, the promotion of same-sex "parenting" is what is *really* a "war on children."

"Telling the Truth Without Redefining Words"

I interact with lots of donor offspring who were told the truth early and often by the people who raised them. Telling donor offspring the truth when they are still young is the right thing to do and seems to be growing in popularity. There is a minor wrinkle to work out, and that is that people raising donor offspring use terminology that makes it unclear what they are telling the truth about.

"The truth about your conception"

No. Their conception story has nothing to do with what they'd believe if they were not told the truth. The truth is that donor offspring are not the biological offspring of the people raising them. The idea in telling the truth is not to mislead them into believing they are related to social parents and their relatives. So telling donor offspring the truth means telling them that they are not the offspring of the people raising them and that in fact they are someone else's offspring. It does also mean telling them that they have other relatives beyond those they are being raised around.

These statements are not true and should be avoided when trying to tell donor offspring the truth:

"Your father and I conceived you with donor sperm."

"I did not reproduce with the donor, I reproduced with your mother."

Or

"I am the biological mother!"

"Donor-conceived individuals"

There is a growing misunderstanding that egg-donor recipients are biological mothers. This is not true.

www.asrm.org/topics/detail.aspx?id=418

www.asrm.org/topics/detail.aspx?id=3634

The good decision to tell the truth often goes hand-in-hand with the use of words intended to conceal the social nature of the parent-child relationship. Truth is undermined by saying it in a way that makes social parenthood sound and feel like biological parenthood. If parenthood does not turn on the reproductive process, then there is no need to try and emulate it when describing social relationships.

"My Story for You..."

I was born in 1968 and raised by my gay mother. My mother was sixteen when she had me. My father was a foreign student, attending college on a student visa. My mother was just discovering homosexuality, so she did not tell my father she was pregnant. She moved away and never told him I had been born. My mother had looked into an abortion. Thank God it was not legal in our state! She kept me and received welfare.

When I was eleven, I asked who my father was. I yearned for my father. My gay mother told me I did not have a father. I persisted and insisted that she tell me the truth. She gave me the only photograph she had of him. She only remembered his first and last name. She said she did not know if he was still in the U.S. or back in South America.

I grew up with an identity problem, which affected my relationships and self esteem, not to mention the indescribable "longing" and constant ache to know the piece of me that was missing.

In my twenties, I searched for my father through private-detective services, contacting the universities he might have attended, writing to the embassy and student-visa departments. I was not able to find him, due to "privacy acts." I prayed to God that I would receive an answered prayer. I never gave up hope to find the father who never even knew that I existed.

Fast forward to present:

I was married three years ago at age forty. I was unable to get pregnant, because my eggs were too old. We consulted with a fertility doctor who advised us to use an egg donor. We looked through the description of young girls who were selling their eggs. One donor girl interested me....She had the same color hair and skin as I. I thought about this idea of using egg donation. My husband and I desired a baby so badly. We were about to do anything to get that little child in to our arms! We prayed to God for guidance. I suddenly thought about the fact that our child may never know his/her biological mother who donated her eggs. I realized I could not bear to allow another child to suffer the same curse of not knowing a biological parent. I knew firsthand about this pain. I had an idea! I called the fertility nurse and begged her to ask this donor girl if she would be willing to give us her identity in case our child wanted to meet her one day. The anonymous donor girl agreed to leave her info with the clinic. Next, the fertility doctor offered to do a split cycle. This meant we could save thousands of dollars if we shared the donor girl's eggs with another local infertile couple like us. This same donor girl actually had a record of previously donating successfully. I suddenly realized this meant our children could have many unknown brothers and sisters in the same city or school, the same age! There were also two other local fertility clinics where this same donor could have also anonymously sold more of her eggs. I worried that one day the descendants of our children may have to ask for a DNA test before dating or marrying. We decided against using egg donation.

My husband and I realized that no matter how badly we wanted a baby, that a baby was not a commodity. These babies will grow up to be real adults…real people. I hope that all of you children grow up and write to your government officials to change the laws regarding anonymous egg/sperm donation in your state. You children do have the power to change the laws so that anonymous egg and sperm donors can no longer be anonymous. (FYI: Adopted children grew up and changed the laws so that adoption records must always remain open in order for them to find their biological parents.) Our state senators and congressmen are available to help! These anonymous donors selling their eggs/sperm should be accountable to meet these children at age eighteen. No one knows the pain and deep longing a child has to know a biological parent. (Only us, the children, will know how this really feels). I am sure when enough of these children from anonymous donors grow up, these laws will change.

With God all things are possible. Nothing is too hard for God. God answered my lifetime prayer to find my biological father! When my husband and I decided to adopt an orphan baby from South America, we received help from a local Peruvian girl who found my father on a social website similar to Facebook! This girl would never have helped me if we had chosen the egg-donation route! Thank God we chose adoption. My father remembered the relationship with my mother. He invited me to visit him. I met he and my brother and sister and large South American Family! I look so much like my father. I now look in the mirror and see my wrinkles, eyes, and hair. I look more like him than his other two children. We also held very similar morals and values. He died of a hereditary illness four months after I found him. I cannot tell you the closure and peace that came from just meeting him. I am a firm believer that with God, anything is possible! I give credit to my faith in God. I feel like a whole pie now. (Before, it was like a piece of me was missing). I am complete.

"Did You Know You Were Supposed to Be Equal?"

There are several countries that allow people to access their donor's identities once they are eighteen years old. The U.S. is not one of those countries, but that's o.k., because what the hell kind of country responds to human-rights violations by asking its disenfranchised people,

"What can we do to make being discriminated against a more pleasant experience for you?"

Donor-conceived individuals in the U.S. will, I hope, begin to compare themselves to those they want to be treated like so they can identify what laws need to change to make that happen.

Did you know that according to Chapter 8 of the Uniform Parentage Act, www.acf.hhs.gov/programs/cse/pubs/2002/reports/essentials/c8.html [link broken], the U.S. Supreme Court decided a series of cases starting in 1968 that precluded discrimination against children whose parents were not married to one another? It's true. And believe it or not, those court decisions culminated in the 1973 National Conference of Commissioners on Uniform State Laws (NCCUSL) approving the original Uniform Parentage Act, which declared,

"Equality for parents and children without regard to the parents' marital status."

I really hope people will stop referring to themselves as donor conceived when what they are, are children of unmarried parents who deserve equal treatment and protection under the law.

Did you know that according to Chapter 1 of the Uniform Parentage Act, www.acf.hhs.gov/programs/cse/pubs/2002/reports/essentials/c1.html#ref_3 [link broken], of the UPA says that is one of the primary reasons they try to identify children's parents is because "Diseases, illnesses, birth defects, and other abnormalities might be passed to children by their parents. Knowledge of both parents' medical histories can help predict susceptibility to certain medical disorders and aid in diagnosis of illness." They say it is to the child's benefit to be in contact with both of their parents.

The UPA also states that identifying a child's parents offers significant social, economic, and medical benefits to children compared to not knowing their identities. They say, "It helps foster in families a sense of parental responsibility, heritage, and self-esteem. Establishing paternity for a child born out of wedlock and having a parent contribute financial assistance for a child's upbringing (that might otherwise come from public funds) benefit the child and society. In addition to providing an alternative source of income for the family, a noncustodial parent could provide a child with access to such "social entitlements" as Social Security benefits, pension benefits, veterans' benefits, and other rights of inheritance." The UPA goes on to say, "Children gain social and psychological advantages from having legally identified parents who are involved in their lives. As well as providing a sense of family heritage, this can be a first step in creating a psychological and social bond between an estranged father and his child."

And the UPA is filled with statements about parentage being conclusively established in contested cases upon testing, so it's clear that not just any old father will do; it has to be their father to warrant recognition as a parent in a contested case.

So far, so great! Things are really looking up for children of unmarried parents; they actually have some rights and people working to identify their parents until you get to the part that specifically excludes children created by artificial reproduction.

The whole intent and purpose of the UPA is undermined by the fact that they do not treat all children of unmarried parents equally such that the entire group can be treated equally with children of married parents. It blatantly creates a subclass of children who have unmarried parents that don't deserve state and federal protection. Those children are noted as being the offspring of gamete donors and surrogacy arrangements, and any child born of an extramarital affair might also fall in with this class of children that does not deserve access to their parents' medical information and does not deserve to foster relationships that foster a sense of heritage.

Furthermore I think that the UPA's noble stated goals of equal rights for children to access vital medical information and foster that sense of heritage should be extended to adopted individuals as well.

It is my opinion that anyone who has been denied access to identifying information about their genetic families needs to begin to look at themselves and their peers in a different light and see themselves as part of this protected class of individuals that deserves access to these simple rights, the same as children born to their married biological parents. It's my opinion that the premise of the UPA with regard to establishing equal rights for children and equal access to support and critical information on their families is the standard to which all other provisions of the act must comply and in turn the standard to which all state laws will be judged against. I think it would be a good idea to stop calling for an end to anonymity and start calling for a start to your rights as United States citizens.

They just need one of you with deep pockets and a law degree to take it up in court, and the law would change.

"My Letter to the Chairman of the Insemination Committee"

[Note from Alana: I received this submission shortly after being interviewed by *Svenska Dagbladet,* a popular newspaper in Sweden. In the interview I revealed I was dating a Swede, whom this contributor refers to as my boyfriend. That Swede is now my husband.]

I typed a letter in 1983 to the chairman of the board of insemination, one of the Swedish parliament's four JO [short for "Justitieombudsmannen," an elected governing official].

It was one of at least 20,000 letters on the same subject, saying in effect the same thing. Not of being inseminated into existence, but of being robbed of the knowledge of who was my real father.

In my case, I received the knowledge at twenty, and had suspected this for ten years, but forced myself not to believe it. My father was then dead, and I have never been able to meet him.

I experienced very, very difficult times ahead and had to leave my university studies then, among other things.

Back to my letter. In 1984 the Swedish parliament passed a law that every child has the right to know his/her parents at latest at eighteen years of age, but the advice is to tell this as soon as possible. So to say, to grow into this knowledge gradually.

The chairman, Thor Swedner (I think) was very attacked for this proposition and this law (it didn't matter at all whom the genes came from, and so forth), and when he defended himself, he used mainly my arguments and sometimes even my own wording.

I felt happy with having achieved something—getting a feedback—to my difficulties before.

Sweden was the first country in the world to adopt this law. Many countries have taken this law and directly duplicating it or taking it and making some changes. Great Britain took this law one or two years ago, for example.

The numbers of artificial inseminations with an anonymous donor has dropped from 3–4,000/year to about 300/year, and is mostly performed in Denmark, as it is forbidden in Sweden.

The law is still in effect in Sweden, even if it is under constant attack. But all those letters, saying the same thing, telling the same story, in different ways, act as a protector of this law.

What I originally wrote I could send you Alana, but you have to ask your boyfriend to translate it to English.

With kind regards

"Phony Bastards and Quasi Marital kids"

I've helped reunite roughly 200 families for free since 1997, so I've seen an awful lot of falsified birth certificates. Turns out that adopted people don't have the market cornered on birth certificates containing false and misleading information. For some, there is no sealed original certificate containing the names of their real parents; for them, there is no record of the truth except the blood in their veins and the unfamiliar face in the mirror. Falsification comes in the form of a certificate that lists no father, as if that were even a remote possibility. Fatherless? Really? No man reproduced to create that child? No man owes that child his support? No man is responsible for raising that child to adulthood? Why not put father unknown so that if he does ever turn up his name can be entered and he can be obligated to support his child? Why should there be a time limit for unmarried men to claim their children? How can children be adopted out from under fathers who did not know of their children until too late, why is it possible to adopt a child without the signed consent of the father and the woman who gave birth. Birth certificate falsification most often starts with the concept of a phony bastard who needs a father. Not their father mind you, because just any father will do.

Every child has a father that creates them and it's him that has a duty to support his child, and nobody should interfere with his performance of that duty. Oh but they do interfere, and when they do, the child ends up not related to the man named as father on their birth certificate. Quasi-marital kids, it's where the stepfather is named on the birth certificates of his wife's children. Quasi-marital children are not adopted by their stepfathers; there is no proof that their own fathers even know they exist, let alone consented to allow their stepfather to raise them. Many people who call themselves donor conceived are just quasi-marital kids, related to the mother on their certificates but not to the father named on their certificates. There is no proof that the man that made them intended to make them; there is no signed consent to show them that their father intended to abandon them to be raised in exile from their paternal families. No proof. None at all. Willing to be known? When? Eighteen years after birth? A donor number? Really? How do they know for sure that donor's DNA and will match the kid holding the ticket with his number on it? What if it doesn't? What then? What if their father is another patient at the same clinic and not some altruistic baby-hander-offer? What if all the profile sheets are nothing more than creative writing by medical secretaries working their way through night school? No proof, no consent. Quasi-marital kids and phony bastards. Remember that. Those kids have no voice, and they need one; they've lost their father's family as well.

Anyone considering starting a family of their own using someone else's genes should understand that you cannot buy claim to someone else's child. It does not matter how early you insert yourself in the process, it's not early enough to make it be you that reproduces and not them. Artificial reproduction does not make artificial children, it makes artificial parents for real children that already have parents. Remember that, too. They will have a sense that they were abandoned even though they did not exist when the sperm or eggs were traded, because motherhood and fatherhood for those people would not start until they were born anyway. They sign papers saying they relinquish their parental rights upon the birth of any child born of their donation. The child's parents still chose not to raise them.

There is not a man on earth that becomes a father when he deposits his sperm; it makes no difference if the deposit is made to a sperm bank or a vagina; men become fathers when their children are born. Women become mothers when their children are born. If you are related to a child as that child's parent, you are that child's parent, and it makes not a bit of difference what the law says or what the people raising your child think. If you are thinking of donating your eggs or your sperm, remember that you would not be a parent until the birth of your child anyway. When your children are born, it won't matter how they came to originate from you. A father is just as much the father of his children if they have the same mother as if they have all different mothers. The tragedy is that this society does not treat the children that way.

I've seen some very happy reunions. People that called themselves donor conceived, that were really just quasi-marital kids, let the donor number fall by the wayside and start referring to their donor as their father, because that is what he always was. Lots of reunions turning fatherless children into the sons and daughters they are. Maybe some people search and get the door slammed in their faces; I've never seen it happen. I've only seen truth bring peace. Sure, over time you might not always get along, but that is family. Good luck finding yours. Truth and peace that is.

"It's Scary to Need a Donor and Read These Stories"

My husband has a genetic condition called Klinefelters Syndrome, or XXY. Contrary to popular belief, this does *not* make him "half female." XXY has a scale. Some have worse symptoms than others. He bears none of the physical characteristics other than long arms and legs and smaller than average testicles.

Due to his parents' lack of care or interest in him, it went undiagnosed until last year, after months of us trying to become pregnant. We went as far as having both his testicles cut open and parts of them removed to see if we could find any sperm at all....There was nothing. Not even one. It broke our hearts.

Using donor sperm is our only option to become parents, to experience pregnancy and birth together. We are already on the outer in regard to being "normal" parents. I have been reading and researching for months. Reading some of these offspring stories scares the pants off me.

I understand that everyone is entitled to know their biological parent; we plan to tell our children from the beginning. He or she will know that it took three people to bring him/her into this world.

It scares me that our child will grow up and never know or possibly even *care* about the hell we went through, because let me tell you, it *is* hell dealing with infertility. You can't know how it feels to imagine yourself dying without the chance to kiss a newborn, to change a stinky bottom, to be greeted in the morning by chubby arms around your neck. Most people have the dream to have children. Everyone wants a child for the same reasons, and they are all selfish reasons. Some people are more blessed than others with the ability to have children naturally.

I hear the argument about "That's what nature intended for you" a lot, but please let me tell you that I am 100 percent healthy and fertile. I have had every test under the sun to make sure of it. It's definitely *not* what nature intended for *me*.

I could walk away from my husband and fall pregnant to another man and never think about infertility again....But what sort of person would that make me? Would I be able to live with myself after that? Would I look at my future children and think, "I broke a man's heart and destroyed his soul for something that I selfishly wanted, and he was interfering with that dream"? Could I live with myself knowing I had walked away from the man I married, the one who I truly want to spend my life with, because he is infertile through no fault of his own (vasectomy, drug abuse, accident), to find a fertile man who I could love but probably always resent because I had to leave the man I married?

As a potential parent of a donor-conceived child, I beg you to try to see it from a parent's point of view. You are absolutely entitled to be annoyed at your parents for hiding it, for not knowing your biological parent, but please know that it can't have been an easy choice to make for them.

"Happily Ever After"

I found my prince charming. I waited and prayed for him for so many years, I kissed "a lot" of frogs, and my heart was broken a few times, but it was worth the wait. His name means "God's gracious gift," and with my faith in God, I know he is. We had a perfect wedding. We live literally in a house with a white picket fence. But after a year of marriage and no baby, we decided to visit the fertility clinic. Before seeing the fertility doctor, we had to attend a fertility class that provided information on fertility issues. The speaker told us statistics on the percentage of couples that are infertile; I didn't think much of it, because I knew it would not be us. She (speaker) told us her story and how her infertility issues lead her to adopt a child. She gave us forms that stated which medical test I needed and which ones my husband needed. We left the meeting but we did not follow up with any of the tests. I just felt that

I was under too much stress from working full-time and attending school and just being a newlywed. I quit school and just enjoyed our first year of marriage.

Three other couple friends told us about their fertility issues, but once again I knew it did not apply to us. They gave us their doctor's name and/or books. I thanked them and put the books in the shelf. After a year of trying to conceive, using all kinds of ovulation kits, we called the doctor and made an appointment. When we met the doctor, he told us how he could help us conceive; the doctor mentioned the tests I needed to undertake to see if I was able to conceive and that my husband had to simply complete a sperm analysis.

A few weeks later, while I was at work, I received a call from my husband. The news was unexpected. I didn't think it applied to us. My legs felt weak, and my heart was palpitating. I felt that life was closing in on me with darkness. As I drove home to be with my husband, many thoughts raced through my mind. My faith in God was being tested; was God punishing me! My fairytale life was ending, and my hopes and dreams of being a mother were being shattered, and the silly thought that I would not experience the first time of taking my child to Disneyland and seeing their joy of meeting Mickey Mouse.

I met my husband at home; we talked, cried, and prayed. I dusted the fertility books from the shelf and read them all. Did I mention that my husband "really, really" wanted to be a father. After reading the books and visiting the doctor, we had only three choices: one is to adopt, two is to use a donor, and the third one is to live childless. The third choice was not what we wanted. The adoption was something to consider as a last resort, but for my story to continue to be "happily ever after," we decided to use a donor.

I read the pros and cons of using a donor and even read scientific articles on the effects of children being born by donors. And what other industrial countries are doing about this issue. I have to say, I was a bit terrified by the negatives stories of donor children. But after coming across this website and reading the stories of "donor-conceived" children and parents, I decided that it is the right choice for us. I thank God for the technological advances available to us; if this were a century ago, we would not have this choice. My husband and I would have lived childless.

To conclude, I recently got inseminated by using an anonymous donor. I prayed to God that I become pregnant and I look forward to being a mommy. Once again thank God for the donors; they are giving infertile couples a chance to be parents.

"A Happy Solution"

I'm a forty-year-old woman who has never found the right partner. I am totally ready to have a child—but it's nearly impossible to meet guys when you already want children so much; it just scares them off. I just can't take my time in a relationship with the baby question so strong in my thoughts.

It has been like that for at least eight years now and recently has got so strong that I have come to the conclusion that, against all my previous ideals of what my life would look like, that if I want a baby, I have to focus on a baby first and a primary relationship second.

As an independent woman, I am used to doing things on my own anyway, which is another predicament whenever I meet a new guy and suddenly have to get used to a new person and all that goes with them, and consider a baby as well—no wonder it's never happened and I have had so many shorter relationships that never seem to progress!

So, I have faced my reality that the baby must come first if I am going to have one at all—but how? I have learned a lot on this website, about the challenges that children of anonymous donors face not knowing their real father, being half of their identity. I thought that if I do go down that track, I would at least like to find a donor that is at least willing to meet their child at some stage, to give them that connection, and to be as honest and upfront as possible about the situation. Thanks to everyone for sharing their stories.

I then thought that maybe there's guys in my life, who may have never settled down and found their ideal relationship, that might feel a part of them is incomplete that they have never had children. Maybe we could do some coparenting; they might play a role like seeing them on weekends or special occasions, a bit like an uncle. I would be fully responsible for the baby, but if they were willing to share their time, provide financial and/or emotional support, then they would have a more active role. I would love my baby to know their father and think if it's someone I know, I can come to a good understanding with them (legally as well, if appropriate).

I would be open to a relationship as a couple only if it worked, as the priority is a harmonious relationship to suit the needs of our child. If either of us wanted to be with other people, that would be fine. I have already run this by my family and friends and have heaps of support and willingness from them to help me raise this child.

I think for me, my ideal relationship has a chance of coming when I have a child and my biological clock is no longer ticking. This child will grow up knowing their father, know they were loved and wanted, and maybe be raised by my ideal partner when they appear. Either way they will be very much loved and wanted and will have great relationships around them, as there won't be pressure on these relationships to be forced to stay together for the child.

I wrote myself a wish card and put some lovely quotes and positive words on it, as well as the type of father I was seeking (someone who cares for me and respects me and would like to play this role) and displayed it on my desk.

The next day one of my male friends, who I have been confiding in, in a deep conversation, said they would be the father!!!

I know he'll be a great dad and look out for both of us. He is excited too and extremely supportive and considerate.

We are going to start trying soon. It's unconventional, but I'm extremely happy and grateful for this outcome. It feels perfect to me, I feel like I have been blessed with him being sent to answer my prayers!

I share my story, as this website helped me, and I hope maybe this inspires someone else too who is seeking their own solution. With all my love, XX

"The Deception"

My sister had twins using donor eggs and her husband's sperm. When she called to tell me she was pregnant, she said she has been having IVF but did not tell me the eggs were from a donor. Something felt "not quite right" about the pregnancy, and I began to suspect donated eggs. My mother confirmed it was true when I spoke to her.

Watching my sister go through pregnancy with another woman's and her husband's children felt perverse and disturbing. My sister continues to deceive me and has never told me that the twins are not genetic relatives. My heart broke when I heard it was true. But even before finding out, although I had been deceived, deep down I could feel there was something wrong. The deception itself is also ineffective, because it does not change the genes of the children.

My sister is not the mother of these children, no matter how mentally distressed she became by her infertility, no matter how much she loves the children, no matter how much she wants it to be true. She has taken on a mothering role, one which she will perform excellently, and she will undoubtedly provide a loving home along with her husband. But still she is an impostor. I believe it is unethical to create a baby to satisfy personal desires, which will then harm the child. Robbing the twins of the opportunity to be raised by their biological mother is not something I feel good about. It is like their maternal identity has been stolen and that of another grafted on. It's a poor substitute, which I can't condone. And it is something those children will feel and will carry with them for their whole lives. All the parenting love in the world will not completely fill that gap that exists in them, as countless adoptees and donor-conceived people will testify.

My relationship with my sister has broken down. The glossy brochures of IVF clinics full of pictures of happy, smiling babies do not want to tell you of their dirty secrets. They don't want you to know how this procedure harms existing families, destroying relationships. The IVF clinic exploited my sister when she was at her most vulnerable and distressed. I would like to see donated eggs and sperm made illegal and the doctors prosecuted.

"Reproductive Choice"

I had myself inseminated following a very painful breakup, upon discovering allegations of incest regarding the man I was dating. Why? I'd cry tears of longing when I'd see mothers together with their children of all ages; I come from a culture in which childlessness is simply not an option. My sister twelve years my junior was pregnant. In summary I heard the clock ticking, and it seemed like there were no appropriate men out there.

Once I had made the decision to inseminate, I resolved to do the best for my kid. For example, I planned never to marry if I had a girl, because I wouldn't want my girl to be exposed to a strange man. I made sure to choose an open donor.

The first time I considered going to a sperm bank, I had just cried and cried. It took me months to get used to the idea, and in hindsight, I never truly did—I guess I was deluding myself. As soon as I left the office after my insemination, I began crying again.

The crying lifted, and I started to get excited. But when I actually discovered signs of pregnancy, I was terrified. I thought of all those internet comments from bitter donor offspring, even though I had made sure to go with an open ID program, which they hadn't. The two things that terrified me most were poverty and hardship as a single parent, which I had dealt with before making the decision, but now, added to that were those accusing internet donor-offspring voices that might one day be pointed at me by my own child.

The next couple of weeks were tortuous. I walked into three different abortion clinics and walked out, until I finally took that pill and miscarried my very planned and expensive but unwanted pregnancy a couple of days later.

I was hit with a tidal wave of grief; it felt like I was drowning. I had always been pro-choice, never viewed abortion as murder, especially one so early as mine when no embryo was yet visible, but never dreamed that I'd actually have an abortion myself. Not guilt, but loss. The person who I hurt was myself.

Before aborting, I figured that if I couldn't take the grief, I could just go get pregnant again. But I haven't done that. I'm in limbo, can't figure out what is the right thing. Nothing has changed; not my culture, not my sister who's now pregnant again, not the biological clock. If anything, three years later, it's ticking even louder, and there are even fewer men.

The only thing that makes me feel better is feeling that I did the right thing, but I don't know that that's true. Life isn't perfect for anybody. I think there's a good possibility that things would have turned out just fine for my kid, as they will if I try again.

But I'm stuck.

Conclusion

Clearly, 3PR is not an ideal family scenario for anyone. It causes serious anguish for *every* group—from social parents to donors to children, and even to extended family. The industry is failing to lead the ethical discussion, so we must charge forward and drag them forward with us. Obviously the industry won't invest in educational campaigns that help people achieve their first-choice family (most often, the traditional nuclear family connected through marriage and biology). They make their money through intervening after tragedy; then they ask their clientele to pretend that nothing is wrong.

Here are the three campaigns we need to shout from the hilltops:

1. **Women, don't wait.** We need to properly educate young women to know the realities of their limited fertility, which begins declining steeply around age thirty. We perhaps should cultivate different expectations regarding higher education for women. Let's start encouraging our daughters to work after high school, save money, and focus on marriage and family while they're young and fertile. And then let's empower them to attend college when their kids reach pre-school age. This way they don't have to choose between motherhood and a career. Let's discourage our daughters from accumulating debt.
2. **Make marriage for life.** Second marriages appear to be plagued by 3PR blues. The connection between children and marriage is not a vestige, and couples who are on their second marriages seek to consummate their new relationship with a child—a symbol of their united bodies and lives. They seek new children even when they have children from previous marriages. Allow me to suggest that a world with less divorce will also be a world with less 3PR. How can we cultivate a culture where marriages are exclusive and forever?
3. **Save the males.** Male fertility is plummeting. And to add insult to injury, there seems to be a troubling dearth of marriageable lads out there. In

achieving our dream families where every man, woman, and child wins, we need to address male infertility by giving our money to doctors so they can *actually* cure infertility. Then we need to give some more money to environmentally friendly organizations and products so we can eliminate the toxicity that is giving us cancers and wreaking havoc on sperm count. Then we need to encourage male culture that makes men into attractive mates. Full-time jobs should be cool; porn and beer…not so much.

If we can repurpose our energy for real cures and prevention, rather than satisfying our expedient urges to procreate, we can put an end to a lot of pain. Billions of dollars are being spent every year to employ *very smart* people in the quest for children. We can get it right, but it's going to mimic the quest of the electric car, I'm afraid. The fertility industry is becoming too much like big oil. It's up to the people to demand health and sustainability. We must demand technologies that work for *everyone*, with special consideration for future generations.

For those of you hurting, you have two wise options: heal or build. Move on, or make it better. If you're not happy with how 3PR is being handled, do something about it.

If you are a U.S. citizen reading this and believe in a ban on anonymity, would you take the time, right now, to write your congressperson and educate him or her about the injustice of donor anonymity?

About Alana

Alana S. Newman grew up in and around San Francisco, California. She is an art-school dropout, a musician, a writer, and recently—a mother. She lives in Brooklyn, New York with her Swedish ex-pat husband and daughter and enjoys live music, reading books about gender dynamics, working out in Prospect Park, and preaching the virtues and structural advantages of the traditional nuclear family.

Alana is also the coauthor of *Happy Couple Creed: How to Make Your Love Last a Lifetime* and the co-creator of Happy Couple Workout. You can visit Happy-CoupleWorkout.com to learn more.

If you need someone to talk to directly about donor-conception, Alana is available privately through Ether and can be reached by dialing: 1-888-MY-ETHER (1-888-69-38437), ext. 04560630.

Find out about everything Alana is up to at www.alananewman.com.